THE DIET CMPASS

Bas Kast was born in 1973, and studied psychology and biology in Germany and the US. He works as a science journalist and author. His publications include *I Do Not Know What I Want* (2012) and *And Suddenly CLICK!* (2015).

Thank you, Ellen, for the inspiration!

THE
DIET
COMPASS

the 12-step guide to science-based nutrition for a healthier and longer life

Bas Kast

TRANSLATED BY DAVID SHAW

SCRIBE

Melbourne • London

Scribe Publications
2 John Street, Clerkenwell, London, WC1N 2ES, United Kingdom
18–20 Edward St, Brunswick, Victoria 3056, Australia
3754 Pleasant Ave, Suite 100, Minneapolis, Minnesota 55409, USA

Originally published in German as *Der Ernährungskompass: Das Fazit aller wissenschaftlichen Studien zum Thema Ernährung* by Bas Kast
First published in English by Scribe 2020
Reprinted 2021

Typeset in 11.5/16.75 pt Adobe Garamond Pro by the publishers.

Printed and bound in the UK by CPI Group (UK) Ltd, Croydon CR0 4YY

Scribe Publications is committed to the sustainable use of natural resources and the use of paper products made responsibly from those resources.

9781912854936 (UK edition)
9781950354290 (US edition)
9781925849844 (Australian edition)
9781925938371 (ebook)

Catalogue records for this book are available from the National Library of Australia and the British Library.

scribepublications.co.uk
scribepublications.com
scribepublications.com.au

CONTENTS

Why I radically changed my diet

The day my heart went on strike

It was one spring evening a couple of years ago, when there was still a glorious freshness in the air as I was just setting off for my usual run, that I realised something was not right. Over the previous few weeks, I had grown almost used to something — a new addition to the usual little aches and pains — which I had not been plagued by before. By now, it had become almost normal to me: shortly after setting off for my run, after the first few strides, I would always get a strange feeling in my chest, as if my heart were skipping a beat.

Nothing serious, just some kind of cardiological hiccup that went away again as quickly as it appeared.

As I continued my run on this particular evening, however, I had barely covered a kilometre when I was violently stopped in my tracks, as if I had run headlong and at full speed into an invisible wall — a wall which brought me to an abrupt and jarring halt. I don't know exactly how to describe this sensation. It feels like a fist of steel grabbing your heart and suddenly squeezing. It hurts, but that isn't the worst part of it by far. The worst, most frightening bit is the overwhelming force with which 'it' bears down on you and brings you to your knees. You stop in your tracks, but not because you think you should take a quick breather; no, you are *forced* to stop. All you can do is stand there, holding your chest, gasping for breath and hoping it will go away soon. And that you will be spared this time and will — somehow — come out of this unharmed.

I have no idea how long I stood there, slightly bent over, hands resting on my thighs, coughing, panting for breath. At some point, I cautiously began to move, making a tentative attempt to break into a trot every so often, before taking another break.

I didn't dare try to run again.

●●●

I love running. Not for health reasons — at least, not when I was younger; quite the opposite, in fact. I was like an alcoholic whose liquor was kilometres. For me, good health was always just a given.

I paid no attention to what I ate. Working as the science editor of the Berlin-based daily newspaper *Der Tagesspiegel*, I was easily able to keep my body going for days on end with just coffee and potato chips. I'm embarrassed about it now, but back then my envious nieces saw me as the uncle who ate chocolate for breakfast and finished off the day with a packet of potato chips washed down with beer. Whenever they came to visit, they would ask me, incredulously, 'Do you *really* eat potato chips for dinner?' — 'Sometimes, yeah!' And why not? I could eat whatever I wanted. I was miraculously fat-resistant.

Then, however, sometime in my mid-30s, I lost my talent for effortless leanness. My body was somehow no longer able to put away all that junk food without it leaving a trace. Although I was still running almost every day, I developed a little belly — or, more accurately, an extremely tenacious spare tyre.

Perhaps it would have been better if I hadn't run. Then I would have got fat faster, and I wouldn't have been able to ignore what I was doing to my body. But as it was, my weight gain was slow and insidious, and I considered myself a fit person. Until that spring evening, when my heart slammed on the emergency brake.

You might think I went home that evening and had a good think about my life, that I was shaken into action by this wake-up call from my body. But at first I did — nothing. I clung to my image of myself as a fat-resistant athlete. My body must be wrong.

Months went by, and I carried on as usual. Just as I had grown accustomed to that skip in my heartbeat while running, I now got used the attacks, which were sometimes quite severe, other times less so. Never again was I able to feel as carefree and liberated while running as I had before. Rather, every time I went for a run, I was constantly waiting for my heart to start rebelling at some point. And I didn't usually have to wait very long.

Then came a period when the attacks would visit me at night while I was sleeping. I would flail about in bed, only half awake, grab my pillow or even my wife in a panic. 'There, there, you just had a bad dream,' she would try to calm me. 'You were having a nightmare.' But I knew, or at least sensed, what the real trouble was.

I can imagine what you're thinking. And yes, of course, I thought about seeing a doctor. I was on the verge more than once — but at the last minute, something inside me always balked at actually going through with it. I have nothing against doctors; when it's absolutely necessary, I am happy and grateful to avail myself of the wonders of modern medicine. Just, not before it's absolutely necessary. The way I see it is this: my health is first and foremost my own responsibility, and only at the point when I am unable to take care of it myself does it become the responsibility of a doctor. Despite that — or perhaps precisely because of that — I now had to do something. Something had to change.

•••

So that's how all this started. The breakdown of my own body, which began far earlier than I had ever imagined it would, forced me to change the way I thought: both generally, about the way I had been living my life up until then, and specifically, about all the junk I had shovelled into my body without a second thought. They say that inside every ageing person is the young person they used to be, wondering what happened. So what had happened? There I was, in my early 40s, recently having become a father to a little boy. Had I brought these premature heart problems on myself? What would become of me if I carried on this way?

It never ceases to amaze me how good we are at ignoring our own faults and weaknesses. How we are able to turn a blind eye even while someone is holding a mirror up to us and practically forcing us to look into it. But at some point, if we're lucky, something happens, something magical — or at least, something that I don't think can be fully explained — and the penny drops. Then you're ready. You are finally ready to make a change. More than ready: you *want* to change.

Although I didn't realise it at the time, that was the point at which I began working on this book, which aims to provide a view of what a healthy diet might look like. A diet that helps stave off, as far as possible, those health problems that so often make our lives a misery. A diet that might even help to slow down the ageing process itself.

Admittedly, for me, personally, the aim at the time was quite different — in the acute situation I was in, I simply wanted to be rid of my heart problems. And so I started doing some research, with one simple question in my mind: what do I need to eat to take the best care of my heart?

I plunged headlong into the complex and fascinating world of nutrition, metabolic biochemistry, nutritional medicine, and, not least of all, geroscience — a rapidly expanding interdisciplinary area of study covering everything to do with ageing, from molecular processes all the way up to the mysterious properties of people who live to 100, 110, or even older while often remaining remarkably fit for their very advanced ages.[1] What's their secret? Why do some people age more slowly than others? Why is it that some remain as fit as a fiddle well into their 60s and 70s, yet others are physical wrecks by the time they're 40? What can we do to slow down our own ageing?

I collected studies like a madman, as if my life depended on it, which, in a way, was the case. I pored over research articles — not out of intellectual curiosity, but for purely existential reasons. Papers began to pile up in my study, my living room, my kitchen. Dozens of them, hundreds, and, eventually, more than a thousand (I stopped counting long ago). Months went by.

A year passed, and then another.

In this way, a world of astonishing, even spectacular, findings opened up to me and changed my life forever. Much of what I thought I knew about weight loss and healthy eating was directly contradicted by the scientific research I was reading. I realised, when it comes to nutrition and diets, there's a teeming mass of myths and folk wisdom out there that can seriously damage our bodies.

A good example of this is the shockingly widespread wave of fatphobia that's been rife in our society since at least the 1980s; even now, we are told by various official health organisations that we should exercise the greatest caution when it comes to consuming fat. This is made all the more fatal by the fact that the warning sounds plausible at first: if you eat fat, you get fat. In addition, we're told, fat blocks our blood vessels, like a fatberg wedged in our drains, and that's what gives you a heart attack. So stay away from fatty meat, full-fat milk, that deadly Greek-style yoghurt, butter, cheese (especially cottage cheese), 'heavy' salad dressings, and the rest. Some otherwise quite creditable cardiologists even warn of the dangers of avocados and those seductive little pellets of pure calories that we know as nuts …

What good have these warnings done us? How much has this demonisation of fat helped us? Has this low-fat cult made us thinner and healthier? No. Quite the opposite, in fact. Indeed, it was not until the advent of this fatphobia that the obesity epidemic we see today really began![2] Nonetheless, many influential organisations, such as the German Nutrition Society (DGE), cling stubbornly to this low-fat dogma.

One fatal side effect of this panic about fat is, and has always been, the fact that anyone who cuts fat out of their diet will inevitably consume something else in its place. This will usually be rapidly digested carbohydrates (often called simply 'fast carbs'), such as white bread, potatoes, and rice, or fat-free — but therefore extremely sugary — highly processed foods. These quickly digested, nutrient-poor carbohydrates are increasingly being found to be among the most phenomenally fattening foods. Some are even more fattening than most types of fat.[3]

We now know that fat doesn't automatically make our bodies fat (although, of course, *some* high-fat snacks, such as the potato chips I used to love so much, can be a not-inconsiderable contributing factor to obesity). What's more, for many people, successful weight loss comes *only after* they start ignoring the 'official' recommendations and *increase* the proportion of fat in their diet (more on this in chapter 5). Thus, for overweight people in particular, certain fatty foods can in fact be helpful in their quest to lose weight!

Several kinds of high-fat food can actually be counted among the most beneficial foodstuffs we can eat, and we should be eating more of them, not less:

- Rather than clogging up our veins and arteries, omega-3 fats — found principally in oily fish such as salmon, herring, and trout, but also in linseeds (flaxseeds) and chia seeds — offer protection from fatal cardiovascular diseases.[4]
- Eating two handfuls of (high-fat) nuts a day will not make you fat, but is more likely to keep you slim, as well as lowering your risk of developing cancer by 15 per cent and reducing the risk of cardiovascular disease by almost 30 per cent. Your chance of dying of diabetes will go down by about 40 per cent, and your risk of mortality due to infection will be 75 per cent lower.[5]
- High-quality olive oils contain substances that inhibit one of the body's critical ageing factors, known as 'mTOR'. In this

way, olive oil may even stop the ageing process and prove to be a kind of anti-ageing medicine (see chapter 8 for more).

•••

We are now bombarded with new pearls of nutritional 'wisdom' on a daily basis, so it's no wonder that it makes little impact on us when the latest health-food fad appears in the media. 'Guaranteed! Achieve your ideal weight in just seven days with these ultimate turbo-fat-burner tricks!' Oh, spare me, please.

Precisely because the vast majority of such diets are nothing but far-fetched quackery, many medical practitioners have also ceased to pay attention to them, and dismiss *all* diets as equally bogus. For decades, more-informed circles have stuck to the same advice, despite the fact that it's of no practical help to most people. Their motto, allegedly the only serious diet formula, is: if you want to lose weight, just *eat less and do more exercise*. This is known as the 'energy balance' principle.

But on closer inspection, this strategy turns out to be a deceptive concept. On a purely logical level, the principle might make sense, in the same way that it makes sense for an alcoholic to drink less. But what good is advice like that to an actual addict? It's not as if he or she doesn't already know it.

The related contention that consuming more calories than you burn will inevitably make you overweight is equally unhelpful. Again, this 'explanation' is factually correct, and about as enlightening as 'explaining' Bill Gates's wealth as being a result of the fact that he made more money than he spent.[6] Yes, he did do that, and in no small amounts. But the question is *how* did he manage that? Or, to return to the issue at hand: *what* exactly is it that leads us to eat more than we burn in our day-to-day lives? And how can we halt this process, and reverse it?

One interesting issue here, for example, is the fact that obesity is often associated with inflammation of the brain. It's as if the brain were 'blocked', like a cold-sufferer's nose, and no longer able to 'smell' the chemical signals sent to the brain by the body when it's full. When this happens, obesity simply leads to even more obesity. Alleviating that inflammation (by eating more anti-inflammatory foodstuffs such as omega-3 fatty acids, for example) also helps patients lose weight. The 'blocked brain' improves, it can once again react to the body's satiety signals, and the hunger pangs decrease.

At first, I was surprised — although it no longer surprises me — that so many of us are so sceptical about the official dietary advice we are given, and prefer to turn to other — unfortunately, usually rather dubious — sources of wisdom. I, too, no longer put my trust in so-called authorities on the subject, but rather rely on the available objective data. This book is a summary of the most important results of that data-gathering process, in which I focus on four central questions:

- What's the most efficient way to lose weight?
- What's the best way to use dietary measures as a protection against illness and disease?
- How can we separate dietary myths from scientific facts?
- Can we outsmart our biological clock and halt the ageing process with a carefully controlled diet?

Compass question no. 1: what's the most efficient way to lose weight?

You might think everything has already been said when it comes to this first question. But in fact, this was one of the areas in which I discovered a plethora of helpful facts and findings that are often overlooked by experts and layfolk alike.

For instance, a large-scale study carried out by researchers at Harvard University a few years ago revealed that some foodstuffs can be remarkably helpful in preventing obesity (see fig. 0.1). Those foods include yoghurt and, yes, nuts, those supposed sources of so many calories. As paradoxical as it might sound, the *more* of those foods we eat, the *less* weight we put on. How do yoghurt and similar foods have this beneficial effect?[7] How can eating *more* of something prevent excessive weight gain? Wouldn't that have to be magic? And while we're on the subject: is going hungry inevitable if we want to lose a couple of those extra kilos? Or is there perhaps another, smarter, way?

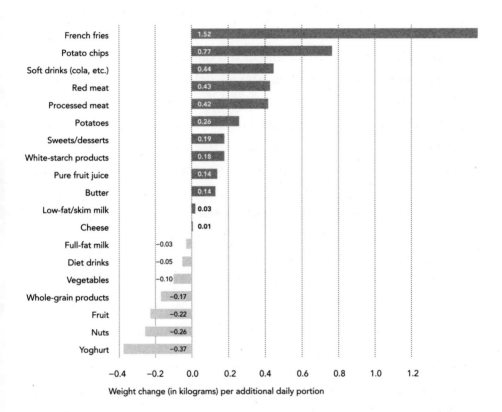

Weight change (in kilograms) per additional daily portion

Fig. 0.1 Fries, potato chips, and soft drinks are associated with particularly large weight gains. Other foods, such as yoghurt and nuts, proved to be a kind of 'slimming food', which helps people maintain their weight. For this study, researchers at Harvard University recorded the weight of thousands of test subjects over a period of four years. On average, the volunteers gained weight in that time. The amount of weight they gained apparently depended on what they ate. An extra portion of fries per day, for example, was associated with a weight *gain* of just over a kilo and a half after four years. Conversely, an extra portion of yoghurt per day was associated with *lower*-than-usual weight gain. 'Processed meat' includes bacon, hot dogs, and similar products. 'White-starch products' covers muffins, bagels, pancakes, waffles, white bread, white rice, and pasta (the question of what exactly starch is, is covered in a later chapter). The 'potatoes' category includes boiled, fried, and mashed potatoes.[8]

In this book, I will examine these questions and many other issues surrounding excess weight and weight loss, including: What are the critical components of a diet that is effective over the long term? Why do diets fail so often? What are the *roots* of that failure? How can we avoid failing when we diet?

The amount of progress we have made in increasing our knowledge in this area is genuinely remarkable. A basic principle has emerged over the years, which helps us understand when we spontaneously stop eating and, contrariwise, under which circumstances we tend to keep on stuffing ourselves in an unchecked, unbridled manner. I believe this principle is extremely important for our understanding of obesity, particularly in our modern world. Anyone who wants to understand their own eating behaviour and lose weight without suffering too much should be familiar with this principle. It's called the 'protein-leverage effect', and I describe it in chapter 1.

On the other hand, it's increasingly clear that there's no such thing as the *one*, single, universal diet that's best for everyone. How well, or otherwise, we respond to a certain dietary approach, such as a low-fat or low-carb diet, depends on our body ('carb' is short for carbohydrate, so a low-carb diet is one that, to a greater or lesser extent, reduces the intake of foods such as sugar, bread, pasta, rice, and potatoes). This is why it's important to test various approaches on yourself and listen to your body rather than sticking doggedly to a pre-planned diet program. I'll take a closer look at this concept, as a minor revolution is taking place here, too: the era of uniform, inflexible dietary guidelines that take no account of an individual's circumstances is over.

In view of the tangled mess of dodgy, not to say daft, dietary ideas and often completely unproven weight-loss advice that's out there, I think it's useful to turn directly to the original academic sources to find verifiable ways of shedding body fat and keeping weight down. 'Losing weight intelligently' is a central concept in *The Diet Compass*, and it's one which I will return to again and again.

Compass question no. 2: how can we prevent the effects of ageing?

Many of the findings I have come across over the months and years turn out to be helpful not just for those who 'simply' want to lose weight, however. No, those research results can also save lives. Preventing illness and remaining fit and healthy in old age through diet is the second main topic of *The Diet Compass*.

Certain diets, for example, can halt the progress of potentially fatal cardiovascular diseases, and even *reverse* their effects. X-ray images allow us to see with our own eyes how even massive blockages in blood vessels can simply vanish.

I'm talking about cardiac patients whose agonies make my own complaints seem harmless by comparison. These are people who were sent home by their cardiologists after a triple bypass operation, with the encouraging advice to buy a rocking chair in which to sit and wait for their imminent demise. Some suffered such unbearable chest pain (angina pectoris) that they couldn't even lie down to sleep and had to spend the nights propped up in a chair. And their pain usually disappeared *completely* in a matter of weeks or months following a change in their diet.[9]

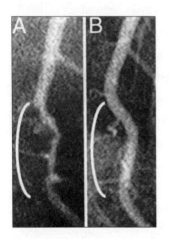

Fig. 0.2 These two X-ray images show a section of a patient's left coronary artery, which is the vessel that supplies much of the heart with blood. The image on the left (A) shows the diseased artery (the white 'tube' which resembles the branch of a tree running vertically). Note the constriction in the area indicated by the white bracket, as if the artery were being squeezed, reducing the flow of blood through it. The image on the right (B) shows the same artery after the patient in this case had followed a strictly vegan, exclusively plant-based diet for 32 months. The constriction has vanished, and blood flow has been restored. The artery now looks completely healthy. The development of vascular disease has not only been completely halted, but has even been reversed, without surgical intervention or medication.[10]

Results such as these are evidence of the power of diet. And this is a power that we can control ourselves; it is — literally — in our own hands. Effects like this show just how fundamentally and profoundly we can change our lives for the better simply by changing our diet.

And these are not just isolated findings. All around the world, scientists are busy exploring ways to treat widespread and/or fatal diseases with special, experimental diets:

- Researchers at the University of Newcastle in England placed test subjects with diabetes (type 2[11]) on a highly restricted diet. Within just one week, the patients' out-of-control fasting blood-sugar level completely returned to normal. After two weeks, they were rid of the 'chronic' condition of diabetes. Since the publication of those results, scientists have almost continuously reported fresh successes using this approach. In plain language: diabetes turns out to be a curable disease.[12]

- The scientist Dale Bredesen, a neurologist and Alzheimer's researcher at the University of California, Los Angeles (UCLA) and former student of the winner of the Nobel Prize for physiology or medicine Stanley Prusiner, currently treats a growing number of patients with memory impairment or those in the early stages of Alzheimer's disease using individually planned diets, supplemented with omega-3 fish-oil capsules, selected plant substances, and vitamins such as D_3 and B. His findings, which are still preliminary but potentially game-changing, are that a majority of patients saw a significant improvement in their memory impairment within three to six months. An initial, small-scale pilot study showed that every patient whose cognitive degeneration had left them unable to continue working recovered so well that they were able to return to work.[13]

My background, as far as my degree and my work as a journalist and author are concerned, is in brain research, so I was particularly impressed when I learned that certain regions of these patients' affected brains were regenerated; indeed, you could even say they 'grew back'. The region in question is the area known as the hippocampus, a structure in the brain that's crucial in memory formation ('hippocampus' is Latin for 'seahorse' and refers to the shape of this brain structure). For one 66-year-old male patient, magnetic-resonance imaging revealed an increase in the volume of the hippocampus after ten months in the order of cubic centimetres: the special diet he was following led to a growth in hippocampal volume from 7.65 cm^3 to 8.3 cm^3![14]

I'm still amazed that I can write sentences like this: cardiovascular disease — the number-one killer worldwide — was not only able to be halted in its progress, but also *reversed*. Really? Diabetes *cured* without any medication at all? The early stages of Alzheimer's might be turned round by means of a diet plan?[15] Surely breakthroughs like this should get about? Diets achieving results that elude the global, high-tech pharmaceutical industry with its multi-billion-dollar budgets should be a hot topic in the print media and internet forums, right? But no, the opposite appears to be the case. Despite, or, more fatally, even *because of* the barrage of sensational headlines about nutrition and diets, most of us have absolutely no idea about this new scientific knowledge. This is a sad fact, and one I hope to change with this book.

Compass question no. 3: how can we separate diet myths from diet facts?

It began with a personal concern. But these and other groundbreaking results prompted me to take my research to a whole new level — I broadened my search. I wanted to find out what the scientific world in general had discovered about the effects of eating a healthy diet. What information lay hidden in the wilderness of diet research, unknown to us despite the fact that it could be crucial for our health and therefore for our lives in general?

Friends and acquaintances began to wonder at the mushrooming piles of papers all over my house (as well as marvelling at my ever-growing library of cookbooks and my often less-than-successful culinary experiments). When I told them some amazing fact I had come across in the course of my research, I was often met with a mixture of fascinated interest and a kind of tedium at having to listen to yet another 'well-meant piece of dietary advice'.

Many people have the impression that the world of nutrition research is, to put it mildly, full of contradictions. Milk is good for you, until it suddenly turns out that milk makes you sick and leads to an early and agonising death, only for milk to be completely and unexpectedly rehabilitated soon after, taking us right back to where we started. Did I do poor butter a disservice when I banned it from my fridge? And what about bread, pasta, potatoes? Is wheat, or rather the gluten it contains, to blame for everything (gluten is a protein found in many types of grain)? Or is it sugar? And then, of course, not to forget the all-important question: is coconut oil the ultimate answer?

Scientific research, with its ever-changing results, is one thing. But we shouldn't forget the what-feels-like at least a million diet gurus out there. It really wouldn't be fair to ignore the unique contribution they make to the confusion, with their sometimes astonishingly bizarre messages of salvation. Every guru has the key to ultimate truth and considers all his or her 'colleagues' to be notoriously dumb. The self-confident low-carb gurus have no time for the boring, killjoy low-fat gurus, and the disdain is definitely mutual. The hipster apostles of veganism seem to be the inverse reincarnation of the hipster apostles of Paleo, hovering round the barbecue trying with missionary zeal to convince us of the merits of eating like a Stone Age hunter-gatherer. And they're all right! Any of them can quote some 'US study' or other that confirms their philosophy! (We will see later how this apparent paradox comes about. And, never fear, there is a way out of this ungodly mess …)

In short, I had kicked a hornet's nest of contradictions. Or, more accurately, I was in the midst of the swarm. What was I to do? I decided that attack was the best form of defence. I resolved to work my way doggedly through this chaos to gain an insight into which of all these contradictory messages were true and which were false. Which philosophies could stand up to such ruthless scrutiny and which could not? Which were myths and which were facts? These are the questions that form the third cardinal point of *The Diet Compass*.

In retrospect, I'm glad I entered this buzzing throng as an outsider, as a science writer whose only skill is to sift through a wealth of academic studies to form an overall view of the story. This position as an outsider turned out to be a great advantage — it allowed me to take an impartial view of all the conflicting claims and often ideologically tinged turf wars. As a diet agnostic, only *one* question was important to me: what really works?

Compass question no. 4: how can you 'eat yourself young'?

How do we even define a 'healthy' diet? (In this book, I use the word 'diet' in the sense of the set of foods a person habitually eats, i.e. as a neutral word, almost as a synonym for 'food', in which losing weight can be, but is not necessarily, an aspect.) As I mentioned at the beginning of this introduction, this book began with my search for a way of eating that would protect my heart. It wasn't until I started researching in earnest that I realised that this book wouldn't only be about finding an answer to that question; indeed, it wouldn't even *primarily* be about that, despite the fact that cardiovascular disease is the number-one cause of death in the world.

I realised that the best and healthiest diet, the one that really deserved the name, would be one that could not only prevent heart conditions, but ideally also prevent as wide a range of diseases as possible. Put bluntly: what use is it to me to have a heart that's in top form, if I'm going to end up with dementia anyway?

So I made it my aim to describe a diet that unites all possible positive health aspects. But I wasn't sure such a diet plan could even be compiled.

As it turned out, achieving this goal was not easy, but certainly doable. To a certain extent, the foods that are good for the heart are generally also good for the brain and the rest of the body as well. Astonishingly, it turns out there's a deeper reason for this connection.

Take a look at the list of the biggest killers in my country, Germany (see fig. 0.3), and you may notice a common denominator, which, at first glance, seems so obvious it hardly deserves attention. Young people's hearts are enviably well supplied with blood. For such people, the risk of suffering a heart attack or a stroke tends towards zero. We have no need to worry about high blood pressure when we are children, let alone such conditions as Alzheimer's disease and other forms of dementia. Our risk of developing cancer also only begins to rise significantly as we age. The same is true of many other diseases, including:

- inflamed joints (rheumatoid arthritis)
- loss of bone mass (osteoporosis)
- age-related macular degeneration (damage to the retina of the eye in the area where vision is sharpest, called the 'macula')
- age-related loss of muscle mass (sarcopenia)
- Parkinson's disease.

It's no coincidence that the most common form of diabetes — type 2 — used to be known as 'age-related diabetes' (as we know, it now strikes more and more adolescents and children, due to poor nutrition and overeating). When it comes to being overweight and that infamous spare tyre of fat, most people share my experience: we don't have to start contending with it until we are past the bloom of youth. The biggest risk factor for all of these conditions is *age*, whatever that means in biological terms.

So an effective strategy would be to draw up a diet plan that targets the ageing process per se and slows it down. Thus, in *The Diet*

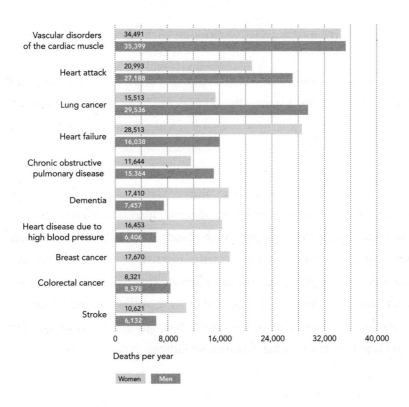

Fig. 0.3 Diet plays an — often crucial — part in all the main killer diseases in affluent countries such as Germany.[16]

Compass, I intend not only to investigate which foods are useful in preventing typical medical conditions associated with ageing, but also to examine — and this constitutes the fourth and final main topic of this book — how diet affects the ageing process overall. Do some foods accelerate the ageing process? And by the same token, and put bluntly, can we 'eat ourselves young', or is that a naive hope?

Just to be clear: it is not my aim to become a frail 180-year-old. This isn't a desperate attempt to add a couple more years on to the end of our lives, come what may.

The aim is a very different one. If you could delay the ageing process, with one stroke you would reduce your risk of developing any of the conditions associated with ageing, from cardiovascular disease to cancer and dementia. Physical and mental decline would be postponed and, ideally, 'compressed' into the final stage of life, rather than wearing on agonisingly for decades.

The key question is not that of *how long* we live, but *how* we age. This is the scenario in my mind — on, let's say, my 88th birthday, I will take my beloved grandchildren for a final swim at the local pool, or — why not, this is my dream after all — go for a final run and then peacefully fall asleep, never to wake again. In the language of medical and public-health experts, this would be the ultimate 'compression of morbidity' (i.e. shortening of the period of life affected by age-related illness).[17]

Okay, maybe I got a bit carried away there. But still, notwithstanding wishful scenarios like mine, scientists are busy deciphering the causes and biological mechanisms of ageing with astounding precision. And their discoveries show that we can directly influence our biological clock via what we eat (or don't eat, i.e. by fasting). The ageing process can be sped up or slowed down depending on what and how (much) we eat. In other words, in a certain sense, you are as young as you eat.

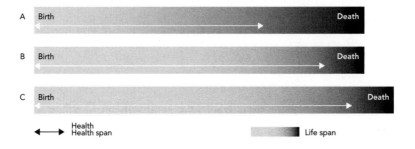

Fig. 0.4 After we reach a certain age, we are often tormented by various ailments (scenario A: years spent in good health are represented in green; those twilight years spent in increasing sickness are represented in grey to black). Even if a healthy diet barely extends our life span, it will still contribute significantly to extending the period we spend enjoying good health (our health span). In this case, the period of ill health is compressed into the final stage of life (scenario B). However, eating a healthy diet can also have both effects, extending both our life span *and* our health span (scenario C). That might sound overly optimistic, but much research, ranging from research on animals to nations with a particularly high life expectancy, indicates that scenario C is not unrealistic.

So, just as an example, it's possible to extend the life span of mice from 100 to 150 weeks by reducing the proportion of protein in their diet from 50 to a level of 15 to 5 per cent. Such Methuselah mice have, among other things, lower blood pressure and cholesterol levels. It's important to note that long-lived animals did not eat less overall — they specifically ate less protein.[18]

Findings like this are relevant, not least of all because the molecular regulators (known by enigmatic abbreviations such as 'mTOR' and 'IGF-1') used by certain nutrients to control the ageing process are remarkably universal across many species, including humans. It's no wonder, then, that scientists recently found something very similar in an investigation involving more than 6,000 people. They discovered that people who eat a lot of (certain kinds of) protein in middle age die earlier; the risk of mortality is raised by 74 per cent, and the risk of developing cancer increases by a factor of four.[19] What kind of proteins were these? Should I steer clear of milk after all? What other nutrients accelerate the ageing process? And how can I turn the tables and influence those molecular regulators positively? You will find out more about all this in the next section.

Summary: what you can expect from *The Diet Compass*

This book is a voyage of discovery into the world of research into nutrition and ageing. It brings together the findings about healthy nutrition gathered over decades of research in laboratories, hospitals, experiments, and observation of people with particularly high life expectancy, to form an overall picture. At its core, it is about the principles behind a healthy diet that can reduce the risk of developing the major geriatric diseases and halt the ageing process itself.

But never fear! *The Diet Compass* will not dictate any rigid eating plans and encourage you to stick to them slavishly. You certainly won't be called on to count calories or any kind of points. You shouldn't be calculating your food; you should be enjoying it.

Rather, the *Compass* provides an overview of which foods we should be eating more of and which we should be avoiding. With that basic framework in mind, individuals can experiment and explore according to their own preferences and tastes. With the background information it compiles, the *Compass* will help you design your own way of eating — on the basis of serious scientific research rather than demoralising weight-loss plans and dietary myths.

It's my hope that you make this book your own, in the best sense of the phrase, and that it not only extends your life in a healthy way, but also improves your day-to-day life. While exploring this topic, I, at least, have developed a love of shopping for food and ingredients I was previously unfamiliar with, and of cooking and trying out new recipes.

Oh yes, and by the way, my heart problems have vanished (as has my spare tyre). I feel fitter than I have for a long time. I can now run as freely and easily as I used to.

CHAPTER 1

Proteins I: the slimming effect of proteins

On the illuminating phenomenon of cannibalism in the Mormon cricket

In 2001, a group of friends from Oxford in England spent a week together in a chalet in the southern Swiss Alps. The gang hadn't travelled to this idyllic part of the world to go hiking or skiing. No, they had come to — eat.

A rich buffet awaited them at their holiday home. It was the beginning of a pilot study that was to become a milestone in the history of dietary research. The results it yielded are of crucial importance to anyone who wants to lose weight efficiently (that is, without going too hungry). That this information has been largely ignored by the general public and most nutrition experts is probably due to the fact that the experiment was thought up by two scientists who don't really belong to the classic fraternity of nutritional medicine. They are Australian entomologists Stephen Simpson and David Raubenheimer. *Insect* researchers? *Really?* What does that have to do with diet? A lot, actually.

So here's the story: Simpson and Raubenheimer had made a strange discovery while observing their insects. The researchers studied the behaviour of the Mormon cricket in great detail, and I will use the example of that swarming insect to illustrate the principle they discovered.[1]

Despite its name, the Mormon cricket, about as big as a human's thumb and dark brown in colour, is actually a katydid. Nevertheless, just like its cousins the true crickets, grasshoppers, and locusts, the Mormon cricket sometimes swarms across the countryside (at a speed of one to two kilometres a day). In this insect's case, that means the grasslands of the western US. But why on earth do they do that? That was the question that interested Simpson and Raubenheimer.

The insect specialists were already aware that the driving force behind this animal's behaviour must be hunger. Unlike those of similar species, the field invasions of Mormon crickets don't leave behind a landscape stripped bare and devoid of all greenery. 'In fact, it is often difficult to tell that a band has just marched through an area,' the researchers were surprised to note.[2] These flightless insects march out in search of food, but mysteriously don't devour the grasses they swarm through. Why not? What are these animals looking for?

The researchers' interest was even more piqued when they observed the crickets in the wild, and found that the creatures are indeed avid, if rather choosy, eaters when they are on the move. For example, they happily consume seed heads, the leaves of legumes, carrion, animal faeces — and *one another*.

The Mormon cricket's cannibalism has almost legendary status among the inhabitants of Utah and Idaho, as it regularly causes traffic problems on the local roads. If a cricket gets run over while crossing a road, compassionate comrades immediately rush to the body of their fallen companion and eat it, only to get flattened themselves by the next set of tyres, which, in turn, attracts another cannibalistic feeding frenzy, and so on until the road is a slippery scene of carnage.

On observing this, the researchers had an idea. And so they designed an experiment to test their suspicion. They set up four Petri dishes of powdered food. The first contained protein; the second, carbohydrates;

and the third, a mixture of the two. The fourth dish was a control, and contained food with no protein or carbohydrates, just cellulose (dietary fibre), vitamins, and salt. The scientists placed the dishes in the path of marching bands of Mormon crickets and watched with interest what happened.

They saw that the crickets were not particularly interested in the dish offering carbohydrates, although they certainly eat carbohydrates in nature. However, the insects headed en masse to the dishes containing protein. Translated into human food terms, we could say they went for the steak rather than the fries.

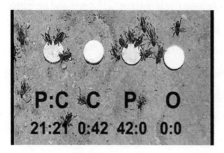

Fig. 1.1 Mormon crickets in search of protein. The dish on the far left (P:C) contained 21 per cent protein and 21 per cent carbohydrate (with the rest made up of cellulose, vitamins, and salt). Dish C contained carbohydrates without protein, and dish P contained protein without carbohydrates (each at a concentration of 42 per cent). Dish O contained only cellulose, vitamins, and salt.[3]

This was confirmation that Simpson and Raubenheimer's suspicion was correct. The Mormon crickets were not driven by simple hunger; they had a specific craving for protein. And what's the tastiest source of protein in a swarm of crickets? Exactly: the other crickets. And that explains why they love to eat each other so much.

The entomologists began to put together the following picture: Mormon crickets converge in large swarms to avoid predation (there's no shortage of animals that are partial to a creepy-crawly protein bar). With their preference for protein, the crickets first head for the richest source in their habitat and devour it until it's gone. That's when the great march begins, as they head off in a desperate search for more protein. What drives the crickets to keep moving isn't just the putative protein source in the distance, but also the protein-hungry cannibal-cricket behind each of them. Either way, the craving for protein proved to be the driving force for their mass march.

On the one hand, the scientists had made a discovery that was not just bizarre, but downright macabre. On the other hand, it turned out that the Mormon cricket isn't as exotic as it first seemed. At least, not as far as its persistent appetite for protein is concerned. A closer look at the scientific research around this reveals that many animals act in a very similar way. Put succinctly, it appears that there is an almost universal 'protein-leverage effect' in operation across all species. It can be briefly described like this: animals are not on an indiscriminate quest for energy, for pure calories. Rather, they feel hungry and continue searching for food until their specific need for protein has been satisfied.

Our diet includes three possible sources of energy: carbohydrates, fat, and protein (alcohol should really be included in this list, too, since it is also a source of energy — while other substances, such as water, salt, or vitamins, for example, are necessary for survival, they do not contain calories that our bodies can burn for energy). Carbohydrates especially, but also many kinds of fat, provide us with energy in the first instance (we will see later that the case of some fats is a little more complicated).

Proteins, on the other hand, are a remarkable special case. Although they are also a source of energy, proteins are primarily used as the building blocks of our bodies, from our musculature to our immune system. This explains one side of the protein-leverage effect: the desperate craving for this very particular building block (not eating protein for a protracted period of time simply leads to death). To describe it in terms of bricks and mortar: when you build a house, you need energy for the machine tools you use; in this case, electrical energy. In principle, you could burn the wooden roof beams or the parquet flooring and use the resulting heat as the energy source you need — however, that wouldn't be very effective, since you need those materials as structural elements of the house you're building. For our bodies, proteins are those structural elements. Bodies can be neither built nor maintained without those basic materials. That means a certain minimum level of protein is essential for life.

And now to the other side. Proteins also have the opposite effect, giving them an additional special status: when an animal has met its need for protein, it tends to stop eating, which is not so much the case for carbohydrates and fat. That means it's much easier to overeat on carbs or fat. Or, in terms of the house-building image: when there's enough building material, the site need not be cluttered with more of it. But you can never really have too much electricity.

From a metabolic point of view, this is connected to the fact that our bodies can't store extra protein as efficiently as they can store carbohydrates and fat. Carbs and fat can be stored in a special form inside our bodies and kept for later. That special storage form is called 'glycogen' (stored carbohydrates) and 'triglycerides' (stored fat). In the broadest sense, we might describe our muscles as a form of protein storage, but, as we know, the muscles of our heart are not exactly eager to be 'burned' to provide the body with energy, although this is what happens in an emergency situation, i.e. when we are literally starving. Under normal circumstances, however, bodies do all they can to avoid 'burning up' their valuable muscles — better to use up their reserves of carbohydrates and fat as a source of energy.

To sum up: in many animals, protein consumption is strictly regulated. They want to eat enough protein, but not too much. The other two main sources of energy — carbohydrates and fat — are secondary influences on an animal's feeding behaviour and the hunger that drives it. Of course, they do have a part to play here, but it's subordinate to the part played by the protein-leverage effect. This effect appears to be common to large swathes of the animal kingdom: it can be seen at work in mice and rats, but also in spiders, fish, birds, pigs, and even non-human primates such as baboons and orangutans. And, who knows, maybe it drives us human primates, too? Do we humans also have a specific craving for protein, and does it control our feeling of hunger and therefore our overall eating behaviour?

By the time Simpson and Raubenheimer got to thinking about that question, they had long been settled at the University of Oxford. One day, the research team happened to meet a bright young zoology student by the name of Rachel Batley. And as luck would have it, Batley's parents happened to own something that would provide a perfect way to investigate the effect of protein leverage on the species *Homo sapiens*: a chalet in the Swiss Alps.

We don't stop eating until our hunger for protein is sated

The buffet in the chalet included everything a hungry heart desires — at least for the first two days. There was muesli and French bread for breakfast, as well as croissants, ham, melon, plums, and many other kinds of fruit. The lunch menu included everything from camembert with fresh bread to tuna, salads, and yoghurt. And dinner was no less sumptuous, with a choice of fish, chicken, couscous, potatoes, and beans; there was also pork, rice, and more kinds of vegetable than you could ever want. There was a large choice of desserts, including almond cake. The guinea pigs — only ten test subjects; as mentioned before, this was an initial, small-scale study — were allowed to eat as much as they wanted. The only condition was that they had to let Rachel Batley weigh their chosen portions and snacks before they ate them, and they could not share or swap any food that had already been weighed. In this way, Batley was able to keep a careful and precise record of who ate how much of what.

The actual experiment didn't begin until the third and fourth days. The participants were divided into two groups — one group was designated as the 'protein rich' group; the other group was 'protein poor'. For the next two days, there were two very different buffets in the chalet. The protein-rich group was only allowed to choose items from buffet no. 1, which contained mainly chicken, pork fillets, ham, salmon and other fish, yoghurt, cheese, and milk, as well as other high-protein foods. Buffet no. 2 had a choice of low-protein items, such as croissants, waffles, pasta, potatoes, couscous, fruit, vegetables, and orange juice. Water was continuously available to all participants. Once again, all participants were allowed to keep on eating until they were full.

For days five and six, the buffets were combined again, and everyone once again had unlimited access to the full range of food available. At the end of those two days, the data-gathering part of the experiment was over.

Simpson and Raubenheimer analysed the data collected so carefully by Batley some time later, during a research residency in Berlin. They found they had gathered the first experimental evidence that human beings behave like migrating Mormon crickets, at least to a certain extent and, usually, in a rather more civilised way. We, too, are driven by the protein-leverage effect: we carry on eating until our hunger for protein is satisfied.

In pure energy terms, human beings require approximately 2,000 to 2,500 calories per day,[4] depending also on body size, physical exercise, age, etc. It's well known that many of us eat far more than that, which is what leads to our being overweight. That's the central doctrine of nutrition research. The rule is: a calorie is a calorie, no matter what food it came from. When we eat more than our bodies need, we put on weight, full stop. Logically, the opposite must also be true: to lose weight, you have to eat less, for example by halving our portions. That is the received wisdom.

However, the experiment in the Swiss chalet revealed that we humans behave in a fundamentally different way — with far-reaching implications for real life, including, for example, attempts to lose weight. This experiment reveals much about why we find it so difficult to 'just eat half' and why that method is ultimately doomed to failure. Although pure energy input is important, food is far more than just a source of energy and, in this sense at least, one calorie is not always one calorie.

Accordingly, the test subjects in the protein-rich group didn't eat approximately the same amount on the days when they were given a high-protein buffet as they did on the days when the buffet was varied. On those days, they consumed 38 per cent *fewer* calories. Their consumption was completely spontaneous, with no one forcing them or even suggesting they eat more or less.

Most remarkable are the results of the analysis of what the subjects actually ate. The smaller number of calories consumed resulted from the fact that the test subjects had unwittingly kept their level of protein consumption *constant*. In other words, those who ate from the high-protein buffet did not stuff themselves endlessly, but stopped eating relatively early. The protein-rich food they were eating meant those subjects satisfied their hunger for protein unusually quickly. The food on the high-protein buffet was apparently so filling that the test subjects put themselves on a 'voluntary diet' without even realising it.

The protein-poor group seems to have experienced the opposite effect. They stuffed themselves and ate 39 per cent *more* calories. I think this finding is so important because it helps to explain why we struggle so much with obesity in our modern times. By overeating, the subjects from the protein-poor groups did nothing different, at a deeper level, to their counterparts in the other group. The data showed that they had also simply tried to maintain their protein consumption at a constant level. However, in order to do so, they had had to stuff themselves as full as possible. Their buffet was so poor in protein that they had no option but to eat far more than they normally would to satisfy their hunger for protein. You could say that their path to consuming a minimum level of protein needed for their bodies to function was strewn with carbs and fats, which they had no choice but to eat.

Seen from the point of view of the classic calorie-counting doctrine, the test subjects behaved in an inexplicably erratic way, with each group acting in the opposite way to the other. Only when seen through the prism of protein leverage does their behaviour no longer seem contradictory, and the behaviour of both groups becomes both explicable and predictable: like other animals, we human beings are not on an indiscriminate quest for pure energy or calories. We're also driven by a craving for a certain amount of protein, and we're remarkably adaptable when it comes to securing our protein quota. If we're presented with protein-rich food, our needs are soon met, we feel full and stop eating of our own accord. If there's too much dilution of protein in our diet, we instinctively eat more — in fact, we keep on eating until our bodies have got what they need. We end up overeating and putting on weight.[5]

That's all very well, but what does it have to do with the widespread increase in obesity levels we see today? And what, in concrete terms, does it mean for our quest to lose weight efficiently?

How modern protein dilution makes us overeat

It turns out that this protein-leverage effect is both good and bad news for us in practical terms. So let's start with the bad news.

According to Germany's National Nutrition Survey, the country's top sources of protein are meat, sausages, milk, cheese, bread, soups, stews, and fish. Men eat an average of 85 grams of protein a day, while women eat 64 grams. That's equivalent to 14 per cent of the energy sources they consume in total (incidentally, a very similar figure to that recorded among the test subjects in the Swiss chalet experiment; the proportion of the energy they consumed that was provided by protein remained consistently between 12 and 14 per cent).[6]

One of my own personal favourite sources of protein is salmon. And salmon is a good example of what lies at the core of the bad news about the protein-leverage effect. The usual salmon fillet you buy at the supermarket or from your local fishmonger has that familiar, distinctive bright-orange hue. And it is just as distinctive for its marbled flesh. That marbling is fat running through the meat. If the fillets you buy look like this, you can be sure you've been sold farmed salmon.

If you have the chance the next time you go grocery shopping, take a look at the wild-salmon fillets for comparison. I say 'if', because wild salmon is much less common and may not be all that easy to find. It's almost impossible to buy fresh. But when you do compare, you'll find the meat ranges from dull pink to dark red in colour and the streaks of fat are much less pronounced, usually almost invisible.

And here's where we get to the crux of the matter: one portion of salmon fillet weighing 100 grams contains 20 grams of protein, irrespective of whether it's farmed or wild. However, the piece of farmed salmon contains 15 grams of fat — *15 times* as much as is contained in a piece of wild salmon of the same weight, which contains only 1 gram of fat (see fig. 1.2)

	Farmed salmon		Wild salmon	
	Weight	Calories	Weight	Calories
Protein	20 g	80 (20 x 4)	21 g	84 (21 x 4)
Fat	15 g	135 (15 x 9)	1 g	9 (1 x 9)
Carbohydrates	0 g	0 (0 x 4)	0 g	0 (0 x 4)
Total		215		93

Fig. **1.2** Farmed and wild salmon contains approximately the same amount of protein, but farmed salmon is much fattier and thus provides less protein in relation to the calories it contains. You might say this means farmed salmon is relatively protein-diluted. The nutrition facts are taken from two salmon fillets on sale at my local supermarket and refer to a 100-gram fillet.

Of course, the rich, 'succulent' fat in farmed salmon really comes into its own on the palate in some dishes — for example, sushi. I occasionally eat sushi, preferably homemade, and there's no health-based reason to stop eating it altogether. However, it's important to bear in mind that farmed salmon is highly protein-diluted in relation to its calorie content. Compared to wild salmon, the farmed variety is so full of fat that it provides far less protein per calorie ingested than a wild salmon (i.e. a natural, 'normal' salmon). You have to eat more calories to get your daily quota of protein. Almost without you knowing it, your body entices you to overeat. It's not your fault. Your body is just doing its job and making sure you survive.

It's important to be clear here: this is not about demonising fat per se. On the contrary, the so-called omega-3 fats contained in salmon are in fact extremely good for our health (and the same is true for other fatty foods such as olive oil, avocados, and nuts — but more about all that in the chapter on fat). The point here is simply that a protein-diluted, farmed salmon is like a kind of Trojan horse, smuggling far more fat, and therefore far more calories, into our bodies than we think. Our bodies crave protein and have us bite into a piece of salmon in joyful anticipation of quelling our hunger for it — and what do they get? A concentrated shot of calories in the form of fat as a bonus!

There are nine calories in a gram of fat, which is more than twice as much energy as is contained in protein or carbohydrates, which each contain about four calories per gram. So, with every mouthful of farmed salmon we eat, we are taking in four times more calories than with a mouthful of the wild variety, without ingesting even one iota more of protein.

Of course, fat also makes us feel full, and, as we'll see, a high-fat diet can help certain people lose weight, if they cut out other food types at the same time. As a whole, we can say that if the phenomenon of protein dilution were restricted to salmon, the problem would not be as bad as it in fact is. However, salmon and the fat it contains is just one among many examples of protein dilution in our modern diets.

Unfortunately, protein dilution runs through our entire modern foodscape. Exactly like salmon, meat, especially processed meat such as sausages, is highly 'fattened'. Wild game meat is typically far, far leaner than the farmed meat we usually eat (roughly 4 grams of fat per 100 grams of meat, compared to 20 grams per 100 grams).[7] Sausage is in a category all of its own — a highly processed form of meat that simply does not occur in the natural world. Sausage principally consists not of protein, but of fat. In fact, sausage isn't really meat at all in the narrowest sense of the word — it is a high-fat industrial product containing traces of protein.

Yet that's not the only problem with protein dilution. It goes far beyond the issue of fat. You might argue that it's perfectly possible nowadays to eat a diet that is low in fat, and that's absolutely correct.

Our supermarkets are full to bursting with low-fat products: fruit yoghurt with 0.1 per cent fat content, fat-free cookies, low-fat this, light that — the low-fat movement has even produced such delicacies as low-fat pizzas and low-fat mayonnaise! So can't we satisfy our need for protein without taking in too much fat by eating these low-fat products?

Yes, we can. However, *low in fat* doesn't necessarily mean *low in calories*. To compensate for the lack of fat — that is, to render the fat-reduced and therefore bland-tasting food edible at all — low-fat products are often enriched with so much sugar that any protein they may contain is once again highly diluted. Just not by fat this time. Protein dilution by carbohydrates is not an improvement on the situation — quite the opposite, in fact. In pure health terms, sugar and rapidly digested carbohydrates are far worse for us than most types of fat. So once again, the problem isn't fat per se, but protein dilution across the board.

In a way, this can be seen as a worldwide experiment, in which we are all eating from a giant buffet that is not necessarily totally lacking in protein, but is relatively low in protein compared to the number of calories it contains. Like the test subjects in the low-protein group in the Swiss-chalet experiment, we eat excessive amounts of fat and carbohydrates in our craving to meet our need for protein. The result is that we end up overeating, even though we don't want to. We overeat fat and carbohydrates in our quest for protein.

When we consider the fact that this overeating is ultimately driven by a deep-rooted instinct for survival (as described above, eating too little protein leads to death), the true gravity of difficulties that we have manoeuvred ourselves into becomes clear. If I diluted the oxygen in the air we breathe, what would you do? Hyperventilate, what else? What else *can* you do in such an emergency situation? Your body can't survive on less than a certain amount of oxygen, so you start to breathe more. Now, what if I start enriching the oxygen-diluted air (proteins) with calories (carbohydrates and fat)? What will happen then? Who'd have thought it: you get fatter. And the reason you get fatter is that you want to stay alive.

And then you can expect to receive the following excellent advice: if you want to get your weight under control, maybe you should simply — *breathe a little less!* We are often told that obesity, weight loss, and staying slim — it's all just a question of self-discipline. You just need to get a grip on yourself. But what do you think is the more realistic scenario? That people suddenly lost the ability to control their eating sometime over the past few decades? Or that something in their environment got skewed and is now leading their natural instincts astray? Somehow, I think the second explanation might be slightly more plausible. And one central part of this distortion — albeit certainly not the only one — is protein dilution.

It's not limited to farmed fish and meat, sausages, and food with added sugar. Rather, as a rule of thumb, you can say: whenever you opt for any processed food, you can assume it will be protein-diluted to some extent (processed products have added fat, sugar, or both).

Some of the industrially produced 'foods' that fill our supermarkets shelves could even be described as real *protein bait*: tricking us with the smell and taste of protein without actually providing any significant amount of protein.[8] This requires a brief explanation.

As mentioned earlier, the study in the Swiss chalet was a very small-scale, ad-hoc experiment; really just a start. However, since then, further data and larger-scale, better-controlled experiments have repeatedly proven the protein-leverage hypothesis.[9] One such experiment showed that the test subjects in its low-protein group, as before, ingested more calories, and they gained those extra calories by intensive snacking on salty, savoury treats between meals. Here, once again, the subjects were not just hungry, they were specifically hungry for protein.[10] That's why they craved savoury food: snacks of this kind — peanuts, almonds, or pistachios — are little pellets of protein. And when our hunger for protein is satisfied, we stop eating. For this reason, if no other, nuts make an excellent snack. It is no coincidence that nuts can help us maintain our weight (see the Harvard study described in the introduction).

Many industrially processed snacks exploit our cravings, cheating us in fiendishly brilliant ways. Of course, I would never accuse such a thoroughly honest business as the food industry of evil intent, but it has undoubtedly developed the skill of 'optimising' some of its products in such a way that they exude the aura of pure protein although they actually contain barely any protein at all. A good example of this is those notorious chicken nuggets. Okay, it's fast food, of course, but most of us at least would assume chicken nuggets are a good source of protein. In truth, however, these deep-fried 'lumps' of chicken consist mainly of fat — almost 60 per cent! A third of the calories they provide come from carbohydrates, and, after that, in third place, this industrially distorted poultry product does contain residual traces of protein.[11]

Or take potato chips, formerly my favourite snack, for another example. And the particularly fallacious flavour: barbecue, which promises our taste buds and our brains a highly concentrated meat product, but actually smuggles almost exclusively carbohydrates and fat into our bodies (see fig. 1.3). We gorge and gorge ourselves on them in the unconscious hope of sating our craving for protein, but all we end up with is a homoeopathically diluted portion of it. And so we keep on gorging.

Nutritional content	28 g contain on average
Energy	628 kJ / 150 kcall
Protein	2 g
Carbohydrates	15 g
of which sugar	1 g
Fat	9 g
of which saturated	1 g
Fibre	2 g
Sodium	125 mg

Fig. 1.3 It's barbecue season! While they may taste of marinated steak, BBQ-flavoured chips provide no more than 4 per cent of their total calories from protein. Protein bait like this leads us to keep on eating and eating because it takes a very, very long time for these snacks to deliver on their promise and actually satisfy our hunger for protein.

Unprocessed or minimally processed foods	Percentage of protein (calories)
Fish and seafood	68.3%
Meat	52.5%
Eggs	36.6%
Milk and yoghurt	28.4%
Pulses (beans, lentils, peas)	25.6%
Vegetables (broccoli, etc.)	24.9%
Pasta	14.2%
Potatoes	10.8%
Average	**27.6%**

Highly processed foods	Percentage of protein (calories)
Instant soups	32.3%
Processed meat products (sausages, etc.)	31.7%
Ready-made pizza	16.6%
Bread	13.6%
Cake, biscuits, etc.	5.8%
Fruit juices and soft drinks	5.4%
Fries, potato chips	5.1%
Ready-to-eat desserts (e.g. pudding)	2.7%
Average	**9.5%**

Fig. 1.4 Compared to natural foods, highly processed industrial foodstuffs are almost all protein-diluted. The food industry's fiendishly clever trickery makes us really tuck in, but *also* keeps us hungry. After all, food that actually satisfies hunger would be bad for business. 'Percentage of protein' means what proportion (in per cent) of the calories (energy) provided by the food comes from protein. The categories are only very roughly delineated (for example, no distinction is made in the analysis of fish and meat between farmed and wild varieties, which, as we have seen, makes a considerable difference — here the rule of thumb is the same: the more natural the food, the less it's protein-diluted). It is perhaps surprising to learn how much protein, proportionally speaking, vegetables contain. Measured by calories, broccoli, for example, consists principally of protein.[122]

In short: you should avoid all this protein bait and any food that was specifically designed to trick your instincts. What that means in daily life, to boil it down to a simple takeaway, is that you should stop eating industrialised food products of any kind. Although it is not always protein bait as such, industrially processed food is certainly *systematically* protein-diluted (see fig. 1.4). Eat real, natural food. The closer it is to its natural state, the better. Or, as the US journalist Michael Pollan so aptly put it, 'Don't eat anything your great-grandmother wouldn't recognise as food.'[133]

Which diets make us *spontaneously* eat less

That was the bad news. Now let's move on to the good news. You can, of course, use the satiating effect of protein to your advantage. There are many popular nutrition plans and diets centred on the protein-leverage effect — flirting with it, sometimes without even realising it.

The protein principle gives some credence, for example, to the Paleo ('Stone Age food') community's preference for steaks and game from grass-fed animals. Quite apart from the fact that such 'natural' meat is better for you, it also provides more protein per calorie because it's less fatty.[14] A true Palaeolithic diet should contain no industrial food, and so protein dilution is not an issue. On the contrary: in practice, those who follow a Paleo diet usually consume more than enough protein. The result: they feel full more quickly.

Those in the low-fat faction approach the issue from a rather different angle. Their aim is to reduce the fat content of our diet across the board. Ultimately, however, this results in something very similar. When the proportion of fat in our diet is reduced in a targeted way, the proportion of protein tends to increase, and we feel full more quickly.

The low-carb league, by contrast, concentrates on that other energy source we can easily end up gorging on in our quest to quell our hunger for protein. Many well-known low-carb diets ('Atkins', 'South Beach', 'Zone', etc.) are also described as high-protein diets since they replace most carbohydrates with sources of protein. That prevents the hunger pangs, and it's one of the reasons for the sustained popularity of the low-carb movement.

By contrast, there is simply no such thing as a low-protein diet, but, if there is one somehow (because, in fact, there is no diet that doesn't exist), it hasn't gained much traction. Now you know why that is. (As described above, industrial food is the perverse exception to this: it could be described as a kind of low-protein diet — the ideal 'diet' from the industry's point of view, as it leaves us chronically hungry for more ...)

Of course, the best diet is one that's easy to stick to and integrate into our daily routine. So while the well-meaning 'just eat half' principle is correct, it's of little productive use. Heck, it's not as if everybody who wants to lose weight doesn't know they might just possibly be eating too much! That it might be a good idea to eat less. It's not as if they won't have already tried to do that! The real question is *how* we can manage to eat less without suffering too much. It's so easily said, but how can it be easily done? *What* should we eat so that we don't *want* to eat so much? The protein principle offers a sound base for answers to these questions. That principle states that the first thing you should make sure of when embarking on a diet is that your protein requirements are well covered. A high-protein diet will keep hunger pangs at bay.

Dozens of scientific investigations in recent years have underpinned this principle. Yet even protein is not a panacea, and, as always with complex issues, obesity and weight loss are not one-dimensional phenomena. There are many other aspects to the issue, from culturally bound habits to the bacteria that live in our guts, as we will see later. Still, if there's one consistent finding to come out of all those (often contradictory) weight-loss studies, it is that 'protein diets' are especially effective because they are practically the only diets that lead us to eat less *spontaneously*. Here are two examples:

In a Danish study, researchers placed 50 overweight test subjects on two different slightly fat-reduced diets (the proportion of energy provided by fat in the diet was just under 30 per cent). The volunteers were told they should not go hungry, and that they could all eat as much as they wanted. The scientists set up a specially built supermarket in their research centre, where the subjects could shop for free.

Apart from the 30-per-cent-fat rule, there was just one other condition, which was monitored and controlled by nutrition experts. The condition was that the subjects in the first group had to limit their protein consumption to 12 per cent of their energy intake, while those in the second group were to make sure protein made up 25 per cent — quite a large proportion — of their overall energy intake.

The study took place over a period of six months. As hoped, the test subjects had shed some of their extra pounds by that time. However, there were dramatic differences between the two groups. While the participants from the moderate-protein-intake group lost 'only' an average of 5.1 kilos, subjects in the high-protein group lost no less than 8.9 kilos on average. Some even shed more than 10 kilos in the six-month period. And while only 9 per cent of the participants in the moderate-protein group managed to lose that much weight, the figure was 35 per cent in the high-protein group. To put it succinctly: eating more protein makes losing weight easier.

More weight lost also means a greater improvement in general health. In the Danish experiment, the waist circumference of the subjects from the high-protein-intake group shrank by an average of 10 centimetres (the figure was 4 centimetres in the other group), and they had reduced their intra-abdominal fat by twice as much as those in the moderate-protein group. Intra-abdominal fat is the hidden fat deposited around organs such as the liver, kidneys, and so on. When a man looks like a pregnant woman in her third trimester, you can be sure he's not exactly lacking in intra-abdominal fat. This fat is highly metabolically active and far more damaging to health than the subcutaneous fat that you can pinch between your fingers (and which is relatively harmless).[15]

Another, more recent study carried out by scientists at the University of Washington in Seattle involved only one group of test subjects, who were placed on different diets sequentially. For the first two weeks of the study, the volunteers followed a standard diet made up of 15 per cent protein, 35 per cent fat, and 50 per cent carbohydrates.

For the next two weeks, some of the fat was replaced with protein (30 per cent protein, 20 per cent fat, and 50 per cent carbohydrate), but the researchers made sure the number of calories the subjects consumed remained the same. As expected, this change led to no weight loss. However, in this high-protein phase of the study, the test subjects already reported feeling far lower levels of hunger. Despite the fact that they were taking in *the same* number of calories, the higher-protein diet left them feeling fuller. Once again, this shows that 'a calorie is a calorie is a calorie' is not the case when it comes to leaving you feeling hungry or full. If the calorie is consumed in the form of protein, it will make the eater feel full faster.

Then came the third and final phase of the study. All volunteers were kept on the high-protein diet, but were allowed to eat as much as they wanted for the next 12 weeks. The astonishing result was that within 24 hours, the participants reduced their daily energy intake by almost 500 calories spontaneously, without being asked to do so. The excess protein the test subjects were eating made them feel so sated that they voluntarily reduced their food intake as soon as they were allowed to do so.

This satiating effect of protein endured through the following weeks until the end of the study. It's no wonder, then, that the test subjects lost an average of just under five kilos (mostly in the form of fat). Surprised by their own findings, the researchers concluded that dietary protein has an almost 'anorexic' effect on dieters.[16]

I picked out two particularly striking studies here as examples — but in terms of their focus and results, they are by no means exceptional. They reflect the general results of current scientific research into obesity. Only recently, an international team of scientists published a systematic review of the findings of 38 studies of this kind, including data from more than 2,300 test subjects. The conclusion they came to is that high-protein diets lead to greater weight loss than low-protein diets.[17]

In view of these very clear and encouraging results, we might say, 'What are we waiting for? If protein's so good at stopping hunger pangs, then bring on the rump steaks and chicken fillets! Give me omelettes, milk, eggs, and all the other protein-packed foods!

Indeed, there are people who have managed to get their weight problem under control — often after years of failed diets — using this kind of strategy. It's possible that some might have saved their own lives this way, even if their eating habits were not 'objectively' (i.e. for most of us) recommendable. I will examine this more closely later in the book.

For now, let's stay with the overall picture. If we broaden the focus beyond just the issue of weight loss and also consider long-term health impacts and ageing, then the effects of unbridled protein intake become far less simple. The crucial catch is that certain proteins, if consumed too excessively, accelerate the ageing process. Thus, these proteins increase the risk of numerous conditions associated with ageing, if consumed to excess. In the next chapter, I'll look at which proteins these are and how we can strike a balance between weight and health. The chapter ends with the 'protein compass needle', which offers an at-a-glance summary of which proteins you should be eating more of, and which you should be avoiding.

CHAPTER 2

Proteins II: the engine of growth and ageing

Did the Atkins diet kill Atkins?

In early 2004, *The Wall Street Journal* gained access to a report by the New York City coroner that was actually supposed to be top secret and not for publication. The document, stamped at the very top with the word CONFIDENTIAL, was dated 17 April 2003. The man who died on that day and was the subject of the coroner's report was Robert Atkins, inventor of the Atkins diet.[1]

Until then, the story had always been that Atkins died at the age of 72 after slipping on the icy New York sidewalk on his way to work one morning and hitting his head. His death was said to have been the consequence of that head injury. Seemingly a random fatal accident.

However, the coroner's report suggested that the story of the accident might not be the whole story. Whatever the truth, the report gave rise to all kinds of speculation. Scribbled by hand in the staccato shorthand of the medical professional, the report included hieroglyphs like MI, CHF, and HTN. MI stands for myocardial infarction; in other words, heart attack. CHF means congestive heart failure,[2] and HTN is the medical abbreviation for hypertension, which is high blood pressure.

Whatever the immediate cause of his fall that day (An icy sidewalk? A fainting fit? Or a heart attack?), Dr Robert Atkins was clearly suffering from serious heart disease. It's also certain that Atkins had suffered a heart attack in 2002, a year before his death, although this was put down to a viral infection.[3]

This reignited an old argument — between ardent Atkins fans and fierce Atkins haters. While disciples of the diet continued to vehemently defend their guru, detractors saw the doctor's death as ultimate proof of their opinion — of course it was the Atkins diet that killed Dr Atkins!

Quite apart from the fact that establishing the truth in this case will always be difficult, it might be legitimate to point out that Atkins' medical history is a private matter for him and his family and therefore none of our business. The counterargument is that his diet is one of the world's best known, with greater global prominence than almost any other.

Robert Atkins also named his diet after himself — he was such a modest guy — and did everything he could to publicise and advocate for it. There are photos of Atkins demonstratively holding frying pans full of crispy-fried bacon and eggs before the camera, or grinning with relish while brandishing a big carving knife next to a huge joint of meat. Atkins didn't just preach the Atkins diet, he *lived* it.

As a doctor, he promised his patients they would not only 'lose weight without hunger', but gain fresh energy, wellbeing, and, most importantly, health. He proclaimed explicitly that his plan would offer protection from the very conditions he turned out to be suffering from himself: hypertension and cardiovascular disease. Furthermore:

[The Atkins diet] can effectively and quickly make a positive impact on many of the most common annoyances that patients reveal in the privacy of their doctor's office. In all my years of practice, I've heard and seen it all: fatigue, irritability, depression, trouble concentrating, headaches, insomnia, dizziness, joint and muscle aches, heartburn, colitis, premenstrual syndrome, water retention and bloating.

Atkins was the embodiment of the low-carb guru. According to his diet plan, those who want to lose weight can feast on as much meat, fish, cheese, sausages, bacon, butter, cream, and eggs as they like, as long as they steer clear of carbohydrates such as sugar, bread, potatoes, pasta, noodles, and rice. In Atkins' program, fat makes up a very large proportion of the sources of energy, but proteins are also a major source, too: around 30 per cent of calorific intake comes from — primarily animal — protein.[4] That's far more protein than the 14 per cent we normally consume. After reading the previous chapter, it should not come as too great a surprise that those proteins might produce a certain sense of satiety in us.

There are good reasons why the Atkins diet and low-carb diets in general are still extremely popular today. Indeed, when it comes to weight loss, Atkins comes off very well in tests, especially for short-term weight loss. Diets differ highly from individual to individual, but, generally speaking, there are very few that can melt away the kilos as fast as the Atkins diet can.

A few years ago, a team of researchers from Harvard University compared the results of 48 reliable diet tests. Their findings were published in the respected US medical journal *JAMA* (*Journal of the American Medical Association*) and can be summarised as follows: slimmers using the Atkins diet lost an average — as mentioned, differences between individual dieters can be considerable[5] — of just over ten kilos in the first six months. Although that's far from revolutionary when compared to the results achieved with competing diets, at least with the Atkins diet, slimmers can eat as much as they want, which is more than can be said of most diet plans.

However, the data also show that this impressive Atkins weight loss doesn't stand the test of time. Subjects typically start gradually putting weight back on after about six months, irrespective of which diet program they follow. They stray more and more from their diet and return to their old eating habits. After a year, their average weight loss is 'only' just over six kilos. At this point, the differences in results between the Atkins diet and other weight-loss plans largely disappear, and, indeed, some low-fat diets are (minimally) more successful by this stage.[6]

Of course, all diet researchers want to work with good, obedient test subjects. But in the real world, it turns out that very few subjects stick perfectly to the guidelines of the diet they've been put on. (This is not surprising when you consider the fact that subjects in a scientific experiment are assigned to a given dietary group arbitrarily, which is why it's called a 'randomised trial'. That means subjects can't choose the diet they're now expected to stick to day in, day out, and if the diet doesn't suit their taste, bad luck.) Most subjects cheat to some extent, either intentionally or unintentionally. And they inevitably become more lax as the weeks and months go on. This occurs in practically all diets, but it's particularly acute in the case of extreme-weight-loss diets, which, for many, includes the Atkins diet.[7]

It is also notable that for most of us, even the Atkins diet, or perhaps the Atkins diet in particular, because it looks like a tasty option, even a real gourmet feast, at first glance, turns out not to be as attractive in the long run. Many people just can't stick to it.

Ironically, that's precisely where the good news lies. Certain high-protein diets (Atkins is just the most famous example) — efficient as they may be at promoting rapid weight loss — are rather unhealthy in the long term, and 'rather unhealthy' can be seen as an understatement here. So in this chapter, I will attempt to show how you can make use of the advantages of protein-heavy diets without also having to accept the negative health effects of the classic Atkins diet.

We need less protein in middle age

Early indications that too much protein can have negative effects on health have been known for decades, and Atkins could have acknowledged their existence if he'd wanted to. For example, as early as the 1970s, a study published in the respected science journal *Nature* revealed that there's a very easy way to turn rats into heart and kidney patients. All it takes is to raise the proportion of protein in their diet to the level recommended by the Atkins diet (see fig. 2.1).[8] In recent years, the long-held suspicion that something similar might be the case for us humans has been corroborated by science.[9]

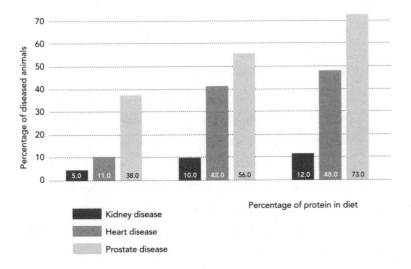

Fig. 2.1 The greater the proportion of protein in the rats' diet, the more likely they were to develop all kinds of conditions.[10]

So protein already had something of a reputation when a series of startling studies from various laboratories around the world revealed in 2014 that some high-protein nutritional approaches can even accelerate the ageing process and shorten overall life expectancy. One of the most significant scientists behind those studies was once again not primarily a nutrition scientist, but a gerontologist at the University of Southern California by the name of Valter Longo. He's the director of his university's Longevity Institute. This is his how he sums up his many years of painstaking research:

We studied simple organisms, mice, and humans and provide convincing evidence that a high-protein diet — particularly if the proteins are derived from animals — is nearly as bad as smoking for your health.[11]

As bad as smoking! Wow! How did the scientist reach this radical conclusion? Is it justified? Valter Longo and his team analysed the nutritional data of almost 6,400 people aged 50 and over. They found that people in middle age, between 50 and 65, who ate a lot of protein (20 per cent or more of their calorific intake, compared to less than 10 per cent) can expect to have a highly increased risk of mortality.

A short note on this: our risk of mortality is, of course, always ultimately 100 per cent. All the longevity studies in the world have been unable to alter that fact. So when you read of a study indicating increased or decreased risk of mortality, it will always refer to the risk of dying within the (limited) observation period of a research project. In other words, not all 6,400 test subjects died within the observation period, of 18 years for the study in question, only a certain proportion of them. Shockingly, the risk was 74 per cent higher for those who ate a lot of protein. What's more, the risk of developing cancer was increased by more than four times — a magnitude that's pretty unusual in nutrition research and usually only turns up in the results of studies on the effects of smoking or alcohol abuse.[12]

But no matter how obvious the association is, a mere correlation is not enough to prove that protein *causes* cancer and other diseases. As scientists say, 'correlation does not imply causation'. It may be the case that, for whatever reason, people who like to eat extremely large amounts of protein have less healthy lifestyles overall than their peers who shy away from too much protein, and that this, rather than the protein itself, is what increases their risk of mortality so greatly.

This correlation vs causation issue is a fundamental problem in many studies of nutrition, and is the basis of much misunderstanding and harm. It's also often at the root of popular myths about nutrition. For example, mistaking correlation for causation was the reason coffee had an undeservedly bad reputation for a long time. There were indications that increased coffee consumption was associated with an increased risk of death — until it was discovered that the opposite is in fact true and drinking coffee lowers the risk of mortality. So what was the problem?

The answer lies in the fact that people who drink a lot of coffee are more likely to be smokers. If this disruptive factor is removed from the statistics, coffee itself turns out to have a protective effect.[13]

Thus, the crucial question is: does eating a high-protein diet actually *cause* cancer? To test this, Longo and his colleagues devised an — in my opinion, ethically dubious — experiment in which they implanted breast-cancer cells into 20,000 female mice and then placed them on different diets. Since cancer cells and microscopically small tumours arise spontaneously all the time in mice (and humans), the critical criterion for disease in the future is how aggressively the cancer can develop (once we reach a certain age, many of us will have mini-tumours lying dormant in our bodies, which, luckily, do not grow so uncontrollably that they end up killing us).[14]

This is where diet plays a critical role, as any nourishment also ultimately nourishes the cancer: *all* the mice who were given a high-protein diet had developed tumours after 18 days. In the low-protein group, however, this 'tumour incidence rate' was only 70 per cent. That means a considerable proportion of tumours could be prevented by eating a reduced-protein diet, *although* the test mice already had a total of 20,000 tumour cells under their skin![15]

This finding makes biological sense when we think about what proteins actually do. Proteins are the body's building blocks. They are the basic materials of cell growth. The cells of our bodies contain special molecules whose job it is to register whether cells can grow or not. One of the central control molecules in this process is called 'mTOR' (mechanistic target of rapamycin).[16] The mTOR molecules lie in wait inside our cells, monitoring the nutrition and energy situation there. If all is good and the cell is getting what it needs, mTOR can confidently give the signal for the cell to grow. With that, our cells get bigger and fatter, and divide and multiply. This is how tissue, for example muscle tissue, grows.

mTOR is primarily stimulated by proteins. No protein, no mTOR activity. That means proteins are the crucial signal substance for cell growth, which isn't surprising, since our cells are mainly made up of protein themselves.[17] And it's no accident that bodybuilders love protein shakes. A simplified formula for this looks as follows, where the arrows stand for 'activate(s)':

Proteins → mTOR → cell growth

While we are still developing during childhood, continuous tissue growth is desirable. But at some point, that development is done. We still require a certain amount of cell growth as adults to replace dead cells, such as skin cells, and to replace other tissue and material. However, that requirement is now no longer as great as it was when we were young. Eagerly wolfing down protein and keeping mTOR running at full power stimulates the body to continue to grow and grow, although the body's cells would rather be taking it easy. Spurred on by all the protein and mTOR, cells continue to build and build, although this building mania is completely incommensurate with the body's need for growth. The products of all that building, which are made up mainly of proteins themselves, begin to accumulate (rather than simply being broken down again), then clump together and begin destroying other cells, for example brain cells, as happens with Alzheimer's disease. You could say cell growth flips to become cell ageing:

Proteins → mTOR → cell ageing

When mice which were already prone to Alzheimer's disease were placed on protein-restricted diet cycles (one week of low-protein food followed by a week of normal food, then another week of low-protein food, and so on), they took longer to develop the disease. One of the reasons for this is that less damaging protein accumulated in their nerve cells.[18]

Now, don't get me wrong here. Neither protein nor mTOR are 'bad'. On the contrary, we couldn't live without either of them. A very low level of mTOR activity is unfavourable over the long term, in particular in muscle cells, which respond by gradually shrinking. However, we overdo it with our western diet and induce our bodies to grow excessively at a time in our lives when such growth causes more harm than good. Rather than more growth, our cells would benefit far more from some maintenance work. For example, it can be extremely beneficial for our cells to take a break every now and then to clear out all the 'building waste' that has accumulated inside them, a process known scientifically as 'autophagy' (self-devouring). When mTOR is active, this beneficial self-cleaning process is blocked.

It is so dangerous for the cells of our body to be stimulated to grow all the time because the favourite activity of every cancer is to grow, and protein building blocks are the main raw material for building new cancer cells.[19] This explains why a diet that's too rich in protein, and the associated hyperactivity of mTOR, creates the ideal conditions for the growth of cancer cells. (Incidentally, there are also substances and foods that actively inhibit mTOR, such as coffee, as previously mentioned, to which its life-lengthening properties can presumably be ascribed,[20] but also substances found in green tea[21] and olive oil — but more on that later.)

Put simply, mTOR acts as a kind of mains switch for the ageing process. However, proteins also drive other signalling substances that stimulate growth, which is why they are known as 'growth factors'. For example, proteins stimulate a critical, insulin-like growth factor known by the abbreviation IGF-1 (insulin-like growth factor 1), as well as the growth-promoting hormone insulin itself.

Later in life, around the age of about 65, our IGF-1 levels plummet sharply.[22] The activation of mTOR also decreases, at least in our muscle cells.[23] Just like all the other cells in our body, muscle cells age. They increasingly wither and die without being sufficiently regenerated and replaced. All this means that ageing is often associated with a gradual process of degenerative muscle loss, and our arms and legs become thinner and weaker. The medical word for this phenomenon is 'sarcopenia'. (Once we reach the age of 40, our bodies lose muscle mass every day, so that in extreme old age we have only about half as much muscle as we did when we were young adults. To counteract this, we should be lifting weights several times a week from the age of 40. That's why I bought a kettle bell — a heavy metal weight shaped like a ball with a handle — which I exercise with regularly.[24])

This may provide the explanation for a surprising phenomenon that Longo and his team also encountered during their data analysis: in our later years, the damaging effect of a protein-heavy diet appears to disappear. In part, the situation can even be turned completely on its head: although a high-protein diet still increases the risk of developing *certain* diseases, such as diabetes, the general effect of eating more protein becomes positive from the age of about 65, when the *overall* risk of mortality is reduced for those who eat a high-protein diet.

And that's not the only positive when it comes to saving protein's reputation. Longo's research team noticed something else interesting: when they restricted their analysis to diets that only include *plant* proteins, the damaging effect disappeared, regardless of age.[25] This finding has since been corroborated by other research groups. According to more recent analyses, eating plant protein could even be associated with a *reduced* mortality risk![26]

For some mysterious reason, our bodies react very differently (even to the point of reacting in the opposite way), depending on whether the protein we eat comes from animals or plants. But why should that be? And what does it mean for our nutrition in practical terms? Are there really no animal proteins that might be good for us? Didn't prehistoric humans eat large amounts of meat?

On the benefits of and misgivings about a Stone Age diet

If an animal, let's say a fly or a rat, is allowed to choose freely between various 'diets' that differ in the proportion of protein and carbohydrate they contain, it will plump not for the mixture that will keep it alive the longest, but for that which maximises its chances of reproducing. In other words, it will choose the diet that increases its likelihood of leaving as many descendants as possible.

A rat, for example, will instinctively make sure it is getting enough protein to grow quickly and be able to reproduce rapidly. The fact that the rat will suffer heart disease and tumours later in life — in the unlikely event that it lives to reach old age — is of secondary importance to nature. The main thing is to leave as many copies of its genes behind as possible.[27]

Female fruit flies live longest on a diet with a protein-to-carbohydrate ratio of 1 to 16 (a portion of protein for every 16 portions of carbohydrate). However, fruit flies lay the most eggs when that ratio is 1 to 4. So which diet will a fruit fly select, given a free choice between more than half a dozen different protein-carbohydrate solutions? You guessed it! She opts for the solution with the 1-to-4 ratio, to maximise her egg production, which she pays for with a lower life expectancy.[28]

Put simply, you could say that those who tuck in to lots of protein are serving evolution. They achieve nature's goal of passing on their genes — albeit by possibly risking their own health in the long term!

I believe this perspective casts an illuminating light on the Paleo diet that's so popular at the moment and which in principle appears recommendable. The Paleo idea is that obesity and disease stem from the fact that we've moved too far away from the diet of prehistoric humans, the diet we were 'designed for'. Those who eat like our ancient, long-gone ancestors, on the other hand, will be blessed with a slender figure and good health ('Paleo' is an abbreviation of the word 'Palaeolithic' — the period of prehistory that began around 2.5 million years ago when human beings first began making stone tools (which is why it is also known as the Stone Age) and ended with the development of agriculture about 10,000 years ago.[29]

There are clear advantages to the Paleo diet, such as the choice of meat it offers, as we have seen. Since there was no industrially produced junk food (cola, potato chips) in the Stone Age, the Paleo diet cuts out any such products, which is of course good both for our figure and our health. So, yes, Paleo *can* be very healthy, when the principle is understood correctly.

The problems begin when it's interpreted as a licence to chow down on as many barbecued steaks as possible — after all, meat made up a large proportion of the diet of Stone Age humans. Yet we will never know precisely what our ancestors ate a million years ago. More importantly: the diet of the average Stone Age human was not necessarily one that led to a long life, because longevity is not the primary 'goal' of evolution. In this respect, nature's goals are very different from our own. Evolution doesn't care whether a fruit fly or a rat live to a ripe old age. It doesn't care whether a man can celebrate his 80th birthday and then go play a last round of tennis with his grandchildren. What evolution cares about is whether we pass on our genes to the next generation — something most people finish doing long before they turn 80. If large amounts of meat and (animal) protein keep us muscular, fit, and fertile when we're young, but cause disease later in life, that's still in line with evolution's goal![30]

I agree with the Paleo movement: yes, meat most probably makes up part of our natural diet. Unfortunately, that says nothing about how much meat is good for us in the long term. It may be that we are doing our genes a great service by eating a meat-heavy diet, while doing ourselves a great disservice in this modern age of longer life expectancies.

Indeed, the vast majority of studies show that eating lots of meat often is bad for our long-term health. That's why it's advisable to exercise restraint when it comes to red meat (beef, pork) and even more so when it comes to industrially processed meat products (e.g. sausages, ham, salami). The latter in particular are by far the unhealthiest sources of protein. So if you're going to give up just one source of meat, you'd be best advised to cut out such artificially deformed food.[31]

To give you an idea of the proportions here: eating 60 grams of processed meat per day (which is equivalent to just one hot dog) increases the risk of mortality by 22 per cent compared to eating 10 grams a day (roughly corresponding to one hot dog per week). For red meat: a daily consumption of 120 grams — a reasonably sized schnitzel — is associated with a 29 per cent increase in the risk of death compared to a consumption of 20 grams a day.[32]

I used to eat meat practically every day. I considered an evening meal without meat as some kind of bad joke, somehow 'missing' something, and a good way of putting me in a bad mood. It took me a long time to change this attitude. Then for some time after that, I saw it as a huge culinary challenge to cook up something spectacular that didn't contain meat. But it can be done! And it can be done better and better with a little practice.

These days, I treat myself, maybe twice a year, to a special piece of grass-fed steak or wild game. Overall, I eat meat once or twice a month, preferably free-range chicken. I strictly avoid factory-farmed 'industrial meat' and processed meat products. The former is partly for emotional and ethical reasons — I now find it difficult to enjoy meat from animals that have suffered. (This is not a judgement about other people's choices; I'm just saying that it's difficult for me to enjoy such meat, although I never gave a thought to it for many years. It simply didn't enter my head that the piece of meat I enjoyed for a couple of minutes came from a lifetime of suffering …)[33]

In short, I haven't given up eating meat entirely, and there's no need to do so from a health point of view. However, I do believe it's a bad thing that we have come to expect meat to appear on the menu every day. I like the concept of the 'Sunday roast'. Anyway, meat is now on my menu only on very special occasions. For example, when my family gathers around the extended dinner table, I like to serve a roast from a farm near where I live, where I know that the animals are kept in humane, free-range conditions and have not been fattened with concentrated feed pellets.[34] As I write these words, I am myself astonished to realise that I hardly miss meat at all now.

The difference between animal and plant protein

Question: where are we to get our protein from, if we eat so little meat? From plants, of course! When we think of protein sources, we automatically think of meat, although many plants and fungi consist to a large extent of protein. Elephants, hippos, and gorillas are all *exclusive* plant eaters (grass, leaves), and it doesn't seem to have a negative effect on their body size or musculature. Good sources of plant protein are: beans, lentils, chickpeas, wheatgerm, oatmeal, bulgur, quinoa, amaranth, seeds — such as linseeds (flaxseeds), chia seeds, sunflower seeds, and pumpkin seeds — and, of course, nuts and peanut butter or almond butter. Vegetables (such as broccoli, spinach, and asparagus) are also often relatively rich in protein.[35]

As already mentioned, plant proteins have proved not only to be harmless, but in fact to have a beneficial effect in protecting against disease. While animal protein is associated, for example, with higher blood pressure and an increased risk of diabetes, plant protein is associated with lower blood pressure and a reduced risk of diabetes.[36] According to a recent, very large-scale Harvard University study, those of us who eat more plant protein — for example, in the form of beans, lentils, and nuts — can even expect to live longer lives.[37]

Scientists don't yet fully understand why plant protein is better for us than its animal counterpart. There are at least two possible explanations. Proteins are made up of building blocks called 'amino acids'. There are 20 amino acids in all. Some of them, such as arginine and glutamine, can be produced by the body (non-essential amino acids). Others, including methionine, leucine, isoleucine, valine, and tryptophan, have to be ingested with our food (essential amino acids). Plant and animal proteins differ in their amino acid profiles. Broadly speaking, animal proteins contain more essential amino acids, such as methionine, while plant proteins contain more non-essential amino acids, such as arginine.[38] And it appears that the effects that are harmful to health and which promote ageing come from an overabundance of essential amino acids. The Alzheimer's mice in the study I described earlier, for example, were intermittently denied essential amino acids, while non-essential amino acids were added to their diet — and were shown to be subsequently less likely to develop Alzheimer's.[39]

Methionine is an important essential amino acid. In fact, methionine is a very special case, because every protein — which is nothing other than a folded-up chain of linked amino acids — starts with methionine. Lack of methionine causes our bodies to stop synthesising proteins. That means the body's building mania is halted.

A tried-and-tested way of increasing the life span of many different animals, such as fruit flies, mice, and rats, is to place them on a low-calorie diet. But it has only been in the last few years that researchers have discovered that such draconian starvation diets are not necessary to achieve this effect: often, it's enough to restrict the animals' intake of *proteins*. (By the same token, the life-prolonging effects of a calorie-restricted diet can be reversed by adding essential amino acids to the animals' diets.) For fruit flies, mice, and rats, all that needs to be done for them to live longer is to restrict their intake of the essential amino acid methionine.[40]

Regardless of the provisional nature of such findings, they have caused some optimists to solemnly preach in favour of a methionine-reduced diet in the hope of cheating the ageing process. Incidentally, this can be easily achieved with a well-planned vegan diet that cuts out not only all animal products, but also Brazil nuts, kidney beans, and many other things.[41] According to the authors of one 'study' on this, one way to further dilute protein intake is to include in your diet 'ample amounts of fruit and wine or beer'.[42] That's certainly an unusual suggestion for a diet to extend your life expectancy!

But seriously, I'm keen to see what scientific findings emerge in the future in this respect. There are indications that some types of cancer like methionine — meaning a methionine-reduced diet may be able to play at least a supporting role in some cancer treatments.[43]

All things considered, however, I think the idea that our entire fate hangs on one amino acid is a little too simple to be true. To pre-empt the arguments here and now: yes, a vegan diet *can* be very healthy — with or without methionine. When applied correctly — that is, if it doesn't consist solely of potato chips and cola — veganism can be one of the healthiest diets available, and I am about to examine why this is so (important: don't forget about vitamin B_{12}! More on this in the chapter on vitamins, chapter 11). From a purely health-based point of view, however, it's not necessary to become vegan, and some studies indicate that there are diet plans that are even a little bit healthier: some animal proteins appear to be rather good for our health. Let's take a look at which proteins those are.

How yoghurt keeps you slim and your body young

Plant and animal proteins differ in more than just the composition of their amino acids. The fact that proteins come in different 'packages' is just as important, if not more so. No matter how sharp a knife we use, there's no way we can cut the proteins out of our schnitzel or our lentil soup.

Meat almost inevitably provides us with a lot of saturated fatty acids, which are all but absent from plants. Meat is also likely to contain more salt, and possibly more iron, than is good for us (in fact, scientists don't currently know what exactly it is about red/processed meat that makes it so harmful to health). By contrast, plants often provide us with abundant amounts of dietary fibre, which may help in inhibiting cancer and slowing the ageing process.

This 'package principle' is also important in the case of the two sources of animal protein I'm about to defend here due to their role in promoting health — and perhaps even rejuvenating us: yoghurt and fish.

Let's begin by looking at yoghurt. This dairy product is more than just gooey milk. It's produced when bacteria are used to ferment milk, which basically means they pre-digest it. The bacteria do this by turning some of the lactose in the milk into lactic acid, at the same time turning it into an almost unique type of food (kefir is similar, and probably has a similarly beneficial effect).[44]

In a series of experiments, researchers at the Massachusetts Institute of Technology (MIT) in Cambridge in the US placed mice on a high-fat, high-sugar diet. The mice not only grew visibly fatter, their intra-abdominal fat swelled up like a balloon. As mentioned earlier, intra-abdominal fat is metabolically harmful fat that accumulates in the abdominal cavity and emits inflammatory substances just like a gland, leading to all kinds of health complications.

Astonishingly, however, this accumulation of belly fat was halted completely when the mice were given some yoghurt to eat along with their fast-food diet.[45] This beneficial effect appears not to come primarily from proteins or other nutrients, such as calcium for example, but from the lactic-acid bacteria in the yoghurt.

The MIT experiments showed that it wasn't even necessary to feed the test mice actual yoghurt along with their fast food to achieve this slimming effect. It was enough to administer the lactic acid bacterium *Lactobacillus reuteri*, which is contained in yoghurt, along with the mice's drinking water. What's particularly astounding is that although the mice who were given the lactobacillus ate *just as much* fast food as those in the control group who didn't receive it, they did not get fat in the way the control mice did. This goes to show once again that losing or maintaining weight is not simply a question of calorific intake, but is also highly dependent on *what* we eat.[46]

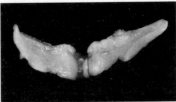

Fig. 2.2 The image on the left shows the abdominal fat of a mouse that was fed a fast-food diet. That on the right shows the abdominal fat of a mouse that received the same amount of fast food, but was also given drinking water containing the lactic-acid bacterium *Lactobacillus reuteri*.[47]

What happens to yoghurt when we eat it? It passes through the stomach into the small and eventually the large intestine, where certain lactic-acid bacteria begin to have a beneficial effect on our immune system and the associated inflammatory processes — not only in the gut itself, but throughout our entire body. The result is that our immune system is stabilised, and inflammatory processes are shut down. That's a welcome development, especially for people who are overweight or old. Although an alert immune system is necessary to ward off infections, if it is in alarm mode for long periods of time, it begins to destroy the body's own tissues.

The details of these interactions are very complicated and are not fully understood. However, it's clear that our immune system ages along with the rest of our body. On the one hand, that means we become more susceptible to infections and less able to cope with them. A bout of the flu or pneumonia can become life-threatening.

On the other hand, a well-functioning immune system must be able to withdraw the body's aggressive defence forces as soon as they've completed their mission and seen off any invaders. It seems we gradually lose this control over our immune system as we age — as if a formerly well-commanded army were descending into anarchy. It's not too much of an exaggeration to say that our bodies are constantly slightly inflamed when we are old.

It's likely that the build-up of old and dead cells and other molecular waste over the years is partly responsible for the fact that our immune system is increasingly mobilised in a desperate attempt to rid the body of this 'old waste'.

However, as we grow older, we usually also grow fatter, and being overweight triggers inflammatory processes. When we put on weight, it is not only that our fat cells multiply; each cell also gets fatter itself. At some point, the fat cells swell up in the restricted space available to them to such an extent that they cut off each other's blood supply. Just like any other cells, fat cells need oxygen to survive, and they can become so swollen that they begin to suffocate. This in turn mobilises the immune system — just as an injury does — to get rid of the damaged material.[48]

Whatever the precise reasons for it, it is typical among overweight people and those who are old to have a continuously high level of inflammatory substances in their body, even when they aren't suffering from an infection such as a cold. The important thing about this is that inflammatory processes worsen almost all the conditions typical of old age, including diabetes, atherosclerosis (the medical name for hardening of the arteries), cancer, and Alzheimer's disease. It's likely that the ageing process itself is driven by inflammatory processes:

Chronic inflammatory processes → ageing/geriatric complaints

All of the above results in the general rule of thumb: foodstuffs that curb inflammatory processes protect our health and may even slow the ageing process. They are also helpful in combating obesity. Many plant substances, such as the curcumin found in turmeric roots or particular substances found in olive oil, as well as omega-3 fatty acids, inhibit inflammatory responses (more on this later) — as do certain lactic-acid bacteria.

In another experiment, the researchers at MIT fed aged mice with yoghurt. They found this made the fur of the mice noticeably more shiny and youthful-looking than that of their age-peers who didn't receive the yoghurt. This is presumably due to the reduced inflammatory processes in their skin. Male mice who received the lactic-acid bacteria didn't experience the testicular atrophy usually associated with ageing. Their testosterone levels remained youthfully high, the mice were slimmer and more active, and they had more muscle mass and thicker fur.[49] (And no, these studies were *not* financed by Danone.)

Yoghurt and the lactic-acid bacteria it contains also has a positive effect on us humans. As already described in the introduction, a large-scale study carried out by Harvard University researchers into the eating habits of more than 120,000 people over several decades revealed yoghurt to be a top-class 'slimming food'. The more of it the subjects ate, the less weight they gained in middle age — a time of life when we usually pile on the pounds. Incidentally, it is not only low-fat yoghurt that has this effect. According to one Spanish study, eating a lot of yoghurt containing a *normal* amount of fat was associated with a reduction in waist circumference.[50] Overall, the fat content of dairy products does not seem to play a pivotal role (for weight and health). Fermentation is far more important!

A number of different experiments have corroborated the theory that lactic-acid bacteria can help with weight loss. In one such experiment, 125 obese men and women were asked to follow a 12-week diet, with half the group also taking lactic-acid bacteria supplements. Over the subsequent 12 weeks, the subjects were allowed to eat normal amounts again. But half the group still received the lactobacillus supplement, while the other half were given a placebo.

The bad news first: the lactic-acid bacteria appeared to make little difference among the male members of the two groups. However, the difference between the women in the two groups was all the greater for it. They not only lost more weight and fatty tissue during the diet phase if they also took the bacteria supplements, they were also far more likely to maintain their lower weight, while the control group gradually put it back on during the second phase of the experiment (see fig. 2.3).[51]

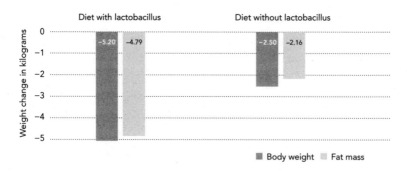

Fig. 2.3 Lactic-acid bacteria can help with weight loss and weight maintenance, at least in women: including lactobacilli in a diet leads to greater loss of weight and fat mass.[52]

There are also preliminary results from research on humans that tally with those of the experiments on mice when it comes to the 'anti-ageing' effect on skin. In one study, 110 volunteers between the ages of 41 and 59 took either a placebo or a lactobacillus 'treatment' every day for 12 weeks. The lactic-acid bacteria appeared not only to result in glossier, more elastic skin, but also to reduce (slightly) the depth of the wrinkles in the volunteers' facial skin (see fig. 2.4).[53]

Fig. 2.4 The photo on the far left shows the skin near the volunteer's right eye at the start of the trial. Progressing from left to right, the images show the same patch of skin after four, eight, and 12 weeks of taking lactobacillus supplements.[54]

I don't want to set too much store by the results of this one study. But as far as the basic approach is concerned, it seems to me that it is more productive to attack the problem of ageing at its root, through our diet, rather than trying to mask the ageing process with all kinds of anti-ageing moisturisers and even the paralysing nerve poison botox. If we can slow the ageing process from the inside, at least we will also be doing our entire body some good, rather than just tinkering with its outer covering.

This is what I do: almost every day — sometimes at lunch, but preferably as dessert in the evenings — I eat a small bowl of unsweetened natural yoghurt with fruit. I usually chuck in a handful of blueberries, strawberries, or raspberries, or a combination of them all, as well as a small spoonful of linseeds or chia seeds, some wheatgerm, and occasionally some pre-cooked porridge, rolled oats, or unsweetened muesli, sometimes with nuts.

Pescetarians live longest

The Japanese of Okinawa are among the most long-lived people in the world, although that applies only to the older generation, who still eat a more-or-less traditional diet. (Members of the younger generation prefer fast food, and there are very clear signs that the extraordinary longevity of the Okinawan population will soon be a thing of the past.) Okinawa is a chain of Japanese islands some three hours' plane journey south-west of the main archipelago. Older Okinawans have always eaten, and still eat, a highly plant-based diet, with occasional small portions of pork. This is a diet that contains only extremely small amounts of fat and protein. At times, the Okinawans have even lived almost exclusively off carbohydrates, in the form of sweet potatoes, which are very popular there. Importantly, however, the extraordinary longevity of the Okinawan Japanese began after World War II, when their sparse diet was supplemented with fish, soybeans, and small portions of meat and dairy products.[55]

Where else might you find such long life? Surprisingly, one such place is the US, the absolute mecca of fast food, which may be home to the people who currently have the longest life expectancy of all. However, the group I'm referring to is not the general population, but the members of a Protestant religious community — the Seventh-day Adventists. You are unlikely to encounter a Seventh-day Adventist in your local branch of McDonald's anytime soon. Followers of this faith see their body as a 'house of God' and treat it with the due respect that implies. (There are Seventh-day Adventist communities all around the world, and it's possible that they have unusually long life expectancies wherever they are. However, I have no relevant statistics to substantiate this theory.) Seventh-day Adventists take regular exercise, almost no one in the community smokes or drinks, and most eat a very healthy diet. Many — although not all of them — are vegetarians. All these behaviours contribute to the fact that Seventh-day Adventists tend to fall prey to disease less often than average, and can expect to live considerably longer than the average American. According to some studies, that extension of life expectancy can be as long as ten years.[56]

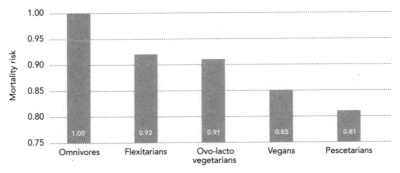

Fig. 2.5 Vegetarian Seventh-day Adventists have a lower risk of mortality than their non-vegetarian peers. Flexitarians are semi-vegetarians — for the purposes of this study defined as those who eat meat at least once a month but no more than once a week, with no limit on their consumption of eggs and dairy products. Ovo-lacto vegetarians are defined here as those who consume eggs or dairy products at least once a month but eat meat (including fish) less than once a month. Vegans: those who consume animal products of any kind less than once a month. Pescetarians: those who eat meat less than once a month, but eat fish at least once a month, with no restriction on their consumption of eggs and dairy products.[57]

Many years of research involving 70,000 Adventists in California resulted in a kind of longevity league table among them. Vegetarians live longer than non-vegetarians, and, among the vegetarians, vegans come off quite well. However, those who topped the table were the pescetarians — vegetarians who eat fish occasionally. Once again, this is only a correlation, but, in this case, I think it's significant, since there appears to be no other fundamental difference between those Adventists who sometimes eat fish (preferably salmon, incidentally[58]) and their peers.

Still, it's a shame that there are almost no experiments that involve placing one group of subjects on a diet that includes fish and another on a similar diet but without the fish to find out whether this has a different impact on the subjects' health. The very few experiments carried out around this issue point to fish being beneficial.[59] Generally speaking, however, we have to rely on observational studies, and such studies suggest that eating fish lowers the risk of disease and death (see fig. 2.6).[60]

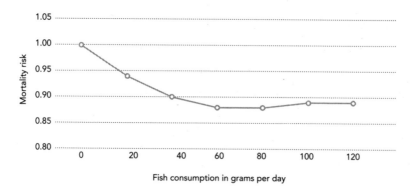

Fig. 2.6 Fish consumption is associated with a lower mortality risk. This graph combines the data from seven observational studies. A brief explanation of how such figures are arrived at: researchers observe a sample group of people, which should be as representative as possible, over a certain period of time (say, a couple of thousand people over a period of several years to several decades). They record the number of non–fish eaters who die during the observation period. That risk is then arbitrarily given the value of 1 (100 per cent). Now that risk can be compared to the mortality risk of those who do eat fish. This reveals that fewer fish eaters died during the observation period. If, as in this case, data is available on how much fish the people ate, the analysis can be further refined, and it becomes possible to state that an average fish consumption of 60 to 80 grams a day is associated with the lowest risk of mortality.[61]

But there are three important caveats here:

- Firstly, the beneficial effects do *not* apply to fried fish, which, for example, is associated with a higher risk of heart failure, while fish prepared in other ways is associated with a lower risk.[62]
- Secondly, this doesn't mean you should gorge yourself on vast amounts of fish. As you can see from the graph in fig. 2.6, the zenith for positive health effects is reached at 60 grams per day. A recent Swedish study based on data from more than 70,000 people placed the optimum amount slightly lower, at 25 (for women) to 30 (for men) grams per day. Indeed, eating more than that resulted in a rise in mortality risk, especially among women.[63] Since fish also contains harmful pollutants, I would recommend restricting your consumption to one or two portions a week. (I tend to eat one to a maximum of two palm-sized pieces of fish, weighing about 100 grams each, per week.)

- Thirdly, this effect depends on the type of fish consumed. One of the pollutants concerned is methylmercury, which mainly accumulates in the bodies of large, long-lived species that are high on the food chain, such as tuna, swordfish, shark, and king mackerel. Pangasius (aka river cobbler, aka basa) — usually hailing from fish farms in Vietnam — contains particularly high levels of mercury as well as other toxins (more on this in chapter 9). Everyone should be cautious about eating such fish, but pregnant women, nursing mothers, and small children should avoid them altogether. Caution should be exercised with common sole (aka Dover sole, aka black sole), too, which can be contaminated with lead. Trout, normal mackerel, and, in particular, salmon, herring, anchovies, shellfish, crab, shrimp, and oysters usually contain only very small amounts of mercury; they can be considered safe for everyone (though pregnant women should avoid shellfish and uncooked oysters as a precaution against possible infections).[64]

Incidentally, autopsies have revealed slightly raised levels of mercury contamination in the brains of fish eaters, which is not particularly comforting news. Remarkably, however — according to one careful study at least — such contamination seems not only to have caused *no* damage, but indeed to have had the opposite effect. *Despite* their increased brain-mercury levels, the fish-lovers showed less evidence of the protein-plaque deposits that are typical of Alzheimer's disease.[65] Overall, fish eaters — just like those who abstain from eating meat — have a measurably larger brain volume. Since gradual cerebral atrophy is a typical phenomenon of ageing, it could be said that the brains of fish eaters stay young for longer.[66]

Another research result that corroborates this indicates that fish eaters experience less memory loss in old age. A long-term study with more than 900 participants whose average age was a little above 80 showed that eating fish once a week is associated with reduced memory loss. That's particularly true of carriers of a gene variant called APOE-ε4, which is associated with an increased risk of developing Alzheimer's disease. According to one calculation, eating fish once a week can reduce the apparent age of a brain by almost 15 years![67]

The protective effect of fish is probably not primarily due to the protein it contains, but rather to its omega-3 fatty acids, which — like yoghurt — have an anti-inflammatory effect. Fish also contains other important substances, including B vitamins and vitamin D. Furthermore, fish is one of the few natural sources of rare but important trace elements such as iodine and selenium.[68]

Finally: in my opinion, fish has one more unbeatable advantage — you really don't have to be a genius in the kitchen to create a fine meal from fish; even I can do it. For example, take a fresh (nice and slimy!) trout, wash it, season it with salt and pepper, and place it in an oven-proof dish. Add a couple of sprigs of rosemary and thyme, a little parsley, or other herbs. If you like, you can add some of your favourite vegetables to the dish. The main thing is that it tastes good in the end. I find tomatoes and zucchinis (courgettes) go well with this fish, as well as onions. Drizzle some high-quality olive oil and some balsamic vinegar over it; add a couple of slices of lemon and garlic, maybe a glug of white wine, and a bay leaf; and put it in the oven (no hotter than 180 degrees). After just 20 minutes, your gala dinner à la *Diet Compass* is ready to enjoy!

Proteins: summary and *Compass* recommendations

Proteins do not only provide energy, they also provide the building blocks for our bodies — for our muscles, blood, bones, immune system, and so on. Many hormones and other messenger substances are also protein molecules. The mTOR molecule is also a protein. We need a critical minimum amount of the 'building material' that is protein in order to survive, and that minimum supply cannot be replaced with other energy sources like carbohydrates or fat. If we do not get this minimum amount in our diet, we, like many other animals, instinctively carry on eating until our hunger for protein is satisfied.

Conversely, our bodies cannot store surplus protein as efficiently as carbohydrates and fat. This explains why our intake of protein is relatively strictly regulated: under normal circumstances, we desire neither too little nor too much protein (the 'protein-leverage effect'). The amount of protein we eat usually fluctuates at around 15 per cent of our total energy intake.

Successful diets usually include large amounts of protein because, of the three main food groups, protein is best at making us feel full. When trying to lose weight, it's therefore important to remember the protein-leverage principle and try experimenting with a protein-rich diet.

However, protein is also an issue for those who want to remain healthy as they age. Diets such as the Atkins diet that rely heavily on (red) meat, ham, sausages, and so on can lead to rapid weight loss. That's encouraging and can be very useful in some individual cases, especially in the short term. In the long term, however, eating too much animal protein of this kind can accelerate the ageing process and increase the risk of developing all kinds of diseases associated with old age.

Similar, or even better, satiation effects[69] — leading to better weight-loss results — can be achieved with very healthy sources of protein. These include fish, shellfish (with poultry as second choice), yoghurt, mushrooms, and plant proteins in general, but, in particular, lentils, beans, broccoli, and other vegetables, as well as seeds and nuts (as a general recommendation: two handfuls of nuts per day — personally, I love nuts so much that I eat a little more than that). For further guidance, see the 'protein compass needle'.[70]

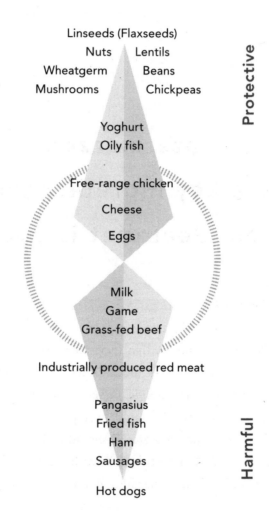

Linseeds (Flaxseeds)
Nuts   Lentils
Wheatgerm   Beans
Mushrooms   Chickpeas

Yoghurt
Oily fish

Free-range chicken

Cheese

Eggs

Milk
Game
Grass-fed beef

Industrially produced red meat

Pangasius
Fried fish
Ham
Sausages

Hot dogs

Protective

Harmful

Protein Compass Needle

The healthiest sources of protein are plants and mushrooms. The best sources of animal protein are yoghurt and fish. For more on pulses (lentils, beans, chickpeas), see chapter 6. It's best to show restraint when it comes to eating industrially produced red meat, pangasius, fried fish, and highly processed meat products, e.g. ham, sausages, and hot dogs.

CHAPTER 3

Intermezzo: the key ingredient of an ideal diet is you

What's more (un)healthy, carbohydrates or fat?

If it's the case that a certain, moderate amount of protein is best for us (not too little, not too much), that inevitably means we can, or even must, tuck in to other kinds of food. Protein usually makes up only about 15 per cent of our calorific intake, so in purely arithmetical terms that leaves 85 per cent of our plate empty. What should we fill it with? Which of the main food groups should make up the difference? Carbohydrates or fat? Which of those is the healthier option?

That sounds like a simple, harmless question. But anyone who asks it should be prepared to face the consequences. It's like diving into the Amazon River filled with masses of underfed piranhas. There is no question that divides people so aggressively as this, and, depending on how you answer it, you will sooner or later find yourself in one of two camps that are deeply hostile to each other.

The answer used to be clear: eat as many carbohydrates as you can and avoid food that's high in fat. This is the position of the traditional low-fat faction. For them, eating fat only makes you fat, because it contains more calories per gram than carbohydrates.

Saturated fatty acids also make you ill by increasing the levels of cholesterol in the blood. I'll take a much closer look at the different kinds of fat later in this book, but, briefly, saturated fats are mainly found in animal products — e.g. meat, sausages, full-fat milk, cheese, butter. These fats do indeed increase blood-cholesterol levels, not least of all the harmful LDL type — but more on that later, too. The cholesterol accumulates in the walls of our arteries (atherosclerosis),[1] which increases the risk of coronary infarction (heart attack) or, when the infarction (blockage) happens in the brain, a stroke (brain attack):

Saturated fats → increased cholesterol levels → infarction

Solution: low-fat diet

The low-fat approach still dominates in the media and enjoys the most 'official' support, for example from the German Nutrition Society (DGE). The consequence of following a low-fat diet is often that we eat more low-fat carbohydrates in the form of bread, pasta, rice, and potatoes. These are generally considered 'staple foods'.

Bread, pasta, rice, and *potatoes?* Sorry, what? To any proponent of the low-carb approach, that sounds like a carefully selected mix of poisons. Low-carb protagonists and other critics of the mainstream position argue as follows: despite the fact that public-health authorities have been warning us urgently since at least the 1980s about the supposed dangers of eating too much fat, and, despite the fact that one low-fat product after the next has taken over our supermarket shelves as a result of those warnings, this has clearly not made us all slimmer and healthier human beings. On the contrary, obesity and diabetes have mushroomed in the same period. It's no wonder, say the low-carb community, since fat, and not least of all saturated fats, have been unfairly demonised. The real danger lurks elsewhere.

The latest incarnation of the low-carb diet is known as LCHF, which stands for 'low-carb-high-fat'. In an LCHF diet, all 'natural' fats are expressly welcome — including butter, cream, cheese, full-fat milk, and oils such as olive and coconut oil. Margarine, on the other hand, as an oil that has undergone an industrial hardening process to make it artificially spreadable, is rejected.

However, the biggest bad guys for the LCHF camp are the carbs. Top of the blacklist is sugar, closely followed by … bread, pasta, rice, and potatoes. Furthermore, the LCHF diet recommends avoiding any vegetables that grow *below* the ground rather than *above* it. 'Below the ground' is synonymous with starch, and starch is a highly concentrated carbohydrate made up of long chains of many sugar (glucose) molecules and therefore not good. This rule means potatoes are taboo, as are carrots, beetroot, and parsnips, for example. Vegetables such as lettuce, any kind of cabbage, tomatoes, broccoli, zucchinis (courgettes), and eggplants (aubergines) are preferred.

What exactly does the low-carb community have against carbohydrates? Three arguments tip the scales for them:

- **Low-carb argument no. 1.** Carbohydrates, and especially the rapidly digested ones such as sugar and soft drinks, but also those staple foods, flood the bloodstream with the monosaccharide glucose (sugar and soft drinks also include fructose, on which more in chapter 4). The pancreas reacts by secreting the hormone insulin, which stimulates the uptake of glucose from the bloodstream by cells. However, when faced with a sugar shock to the body, the pancreas sometimes tries to prevent this by pumping out so much insulin that blood-sugar levels fall so far as to cause a state of hypoglycaemia. When that happens, we need to eat something, and quick. And not just anything, but something that will return the blood's sugar levels to normal as quickly as possible. In other words, what we need is, ironically enough, rapidly digested carbohydrates! In this way, we spend the whole day lurching from sugar rush to snack attack and back.

- **Low-carb argument no. 2.** In addition to promoting the uptake of blood sugar by cells, insulin is also a fat-storing hormone. In other words, carbohydrates cause insulin to be secreted into the blood, which in turn leads to fat being stored. When blood-insulin levels are high, fat cannot be burned — it becomes hormonally impossible to lose weight.

- **Low-carb argument no. 3.** If that wasn't already enough, repeated spikes in blood sugar and insulin levels are quite simply harmful and accelerate the ageing process and the health problems that come with it — ranging from diabetes to cancer.

Carbohydrates → spikes in blood sugar and insulin levels → fat storage/geriatric diseases

Solution: low-carb diet

The above is merely a rough outline of the two opposing positions to provide a general orientation. The details are so complex that I need to explain them progressively, one at a time. But before I move on to those details, let's turn briefly to another question: mightn't we expect this issue to have been resolved, after decades of scientific research have been dedicated to it? Surely it can't be so difficult to reach a conclusion about which of these two opposing schools of thought is correct, can it? However, the fact is that as soon as you begin to compare the pros and cons of each standpoint, it becomes clear that reaching an unambiguous verdict is extremely difficult, if not completely impossible. Strange, isn't it? Astonishing, even. That there can be two so diametrically opposed camps, both of which can be supported with thoroughly reasonable arguments and evidence — from biochemical processes to individual case studies and systematically planned experiments. Where does this intractable contradiction come from?

To put it another way: how probable is it that every proponent of the traditional low-fat position has been wrong for decades? Conversely, how likely is it that every one of the very sizeable number of that position's critics — who include members of prestigious universities and research institutes — is a charlatan or a fool? And if neither of those turn out to be true, then where does our intractable contradiction leave us? If it can't be resolved, can it at least be mitigated?

This dichotomy plagued me for months. The main aim of *The Diet Compass* is to compile from all the research findings and various diet concepts an eating plan that unites all the positive health aspects without adhering to any camp or ideology. For a long time, I assumed there would turn out to be *one* diet plan that would be worthy of being called the ideal diet, an optimum eating strategy that would meet our bodies' needs better than any other. It would then follow logically that there was an optimum amount of carbohydrates and fat. What's better for our health, a low-fat-high-carb or a low-carb-high-fat diet? LFHC or LCHF, that is the question!

For some time, those eight, or rather four, letters almost drove me crazy. After agonising back and forth, however, I gradually came to a realisation. A conviction that my basic assumption was wrong. The deeper I delved, the clearer it became that any attempt to define *one* perfect diet for everyone is not only impossible, but, in fact, counterproductive — especially as far as the relative proportion of carbohydrates and fats is concerned.

There are two reasons for this. Firstly, it turns out that the question whether carbohydrates are unhealthier than fats, or the other way round, is not the crux of the matter. Rather, the crucial issue is the *type* of carbohydrates or fats. Quality is far more important than quantity. *Some* carbohydrates and *some* fats are healthy, others not so much. So the line isn't between carbohydrates and fats. Such a line is artificial and unproductive. To a certain degree, this overarching principle applies to us all.

Yet there is one important exception: recent research indicates that we vary in the amount of carbohydrate we can tolerate. Some people have a metabolic problem with carbohydrates, and their numbers appear to be growing. Such people suffer from a kind of 'carbohydrate intolerance'. For them, a reduced-carb, high-fat diet is to be recommended. What's more, discovering low-carb eating is often a revelation for such people after years of frustration — a kind of liberation. They can satisfy their hunger at last, those extra kilos finally fall away as soon as they change their eating habits, and they rapidly feel much better.

If you happen to be one of those who are intolerant to carbohydrates, then it is not only the quality of the carbs and fats that are important, but also their relative quantities in your diet. In a nutshell, low-carb-high-fat is the best option for you (I examine this particular case in more detail in chapter 5).

For now, let's stick with the bigger picture. The fact that some people's bodies only respond favourably to a low-carb diet might cast an interesting light on the long-running dispute between the low-carb and low-fat factions: why is the low-carb movement so stubbornly opposed to the mainstream (low-fat) position, with their opposition always taking on a new guise? The answer is that it is because bread, pasta, rice, and potatoes actually can become a toxic mix. Not for everyone, for sure, and probably not for most of us, in fact (which is why it's the traditional view). But it is the case for a certain group of people, for some of whom the situation is aggravated by other circumstances. Again, I'll discuss later in this book what those circumstances are and whether you are one of the people whose bodies can't tolerate too many carbohydrates. In this chapter, I want to take a slightly closer look at the overall situation and outline how both a high-carb diet and a high-fat diet can be very healthy, in principle. Let's start with the carbs, before we turn to the fats.

Lots of carbs: from Okinawa to the Adventists

In recent years especially, damning carbohydrates lock, stock, and barrel has almost become a national sport. My advice is to remember the diet of the elderly Okinawans whenever talk turns to 'bad carbs'. The Okinawans are among the healthiest people in the world — and what do they eat? Mainly carbohydrates! Carbohydrates used to account for no less than *85 per cent* of the calories in their diet. Although it has changed over the decades, that proportion is still high today, standing at almost 60 per cent.

In the traditional Okinawan diet, only 6 per cent of the energy came from fats. That figure really bears thinking about: 'low-fat' is usually applied to a diet in which the proportion of fat is 30 per cent or less. So to describe the traditional Okinawan diet as 'low-fat' would be a massive understatement: it is (or was) a 'very low-fat' diet.

And it really doesn't appear to have done the Okinawans any harm. Indeed, Okinawa islanders of the older generation don't just live to an unusually great age, they also suffer considerably less from cardiovascular disease, diabetes, cancer, and dementia than the likes of us. There are around 50 centenarians in every 100,000 Okinawans — more than twice as many as there are in industrialised countries (Germany currently has 'only' 22 centenarians per 100,000 inhabitants).[2] In other words, carbohydrates cannot per se be so terribly toxic.

And here comes the big 'but': on closer inspection, it turns out that Okinawa is a very special case, which can only be applied in very limited terms to the situation we live in. As mentioned previously, Okinawans used to eat very little overall. Their aversion to overeating goes back to a Confucian teaching called *hara hachi bu*, which can be loosely translated as 'eat until you are 80 per cent full'. The Okinawans' good health and longevity could well be due to this 'calorie restriction'. That would be my best guess, in view of the fact that dietary restriction is one of the best ways of extending the life span of a wide range of organisms and animal species — yeasts, worms, flies, fish, mice, and even monkeys.[3]

Ultimately, the crucial factor for the longevity of the traditional Okinawa islanders is simply not known. We don't have that knowledge. This is compounded by the fact that the Okinawans' lifestyle and culture are so radically different from our own, not to mention the genetic differences, and it becomes legitimate to ask to what extent they might serve as a realistic model for us. I think they can do so only in a modest way, at least as far as the relative proportions of food from the main food groups is concerned.

The same applies to many other remarkably healthy ethnic groups that have attracted the interest of researchers in recent years, such as the Tsimané. They are hunter-gatherers who live on the banks of a tributary of the Amazon River in Bolivia. Atherosclerosis is virtually unknown among the Tsimané, a fact that's equally astonishing and encouraging. It could mean that the crucial factor for this leading killer disease is largely self-inflicted, and therefore avoidable. In other words, atherosclerosis is probably not an inevitable consequence of ageing, although that's usually the explanation we're given for the condition.

The diet of the Tsimané consists of no less than 72 per cent carbohydrates, while 14 per cent of their calories come from fat, and another 14 per cent from protein. Their diet is predominantly vegetarian. Is it the Tsimané's diet that protects their hearts from disease? Maybe. Or it could be their lifestyle as a whole. The Tsimané live in simple thatched huts without running water or electricity. Hunting — sometimes still with bows and arrows — can take eight hours or longer and can involve the Tsimané trekking up to 18 kilometres through the rainforest. That means they literally spend all day on their feet, spending less than 10 per cent of the time sitting.[4] Therefore, the lesson we might learn from this is not that eating lots of carbohydrates is good for your health, but that eating a natural diet consisting principally of (high-carb) vegetables, combined with lots and lots of physical exercise, is extremely good for your health.

According to Germany's National Nutrition Survey, carbohydrates account for less than 50 per cent of Germans' calorific intake (approximately 47 per cent), while fats account for 36 per cent. That means the German diet is much less carb-heavy and more fat-heavy than that of either the Okinawans or the Tsimané.[5] This could lead us to conclude that Germans and those from similarly affluent countries should eat less fat and more carbohydrates. And that is precisely the message propagated by the German Nutrition Society (DGE) and other proponents of the low-fat approach. The DGE guidelines state that at least 50 per cent of the calories we ingest should come from carbohydrates.[6] And there's no doubt that this recommendation *can* lead to a very healthy diet.

It can, but it doesn't *have to*. Firstly, it doesn't necessarily follow that more carbohydrates make for a healthier diet — it's no secret that sugar is a carbohydrate and not a beneficial vitamin. Secondly, and more importantly (since the DGE does not, of course, advise eating more sugar, although there are unfortunately some low-fat fans and gurus of veganism[7] who play down the dangers of sugar), there are certainly high-fat diets that have repeatedly been shown in studies to be extremely healthy. Incidentally, the best known of these is not only world famous, but also considered by many leading nutrition scientists — including Walter Willett of Harvard University[8] — to be the ultimate healthy diet: the so-called Mediterranean diet. Depending on how it's followed, the proportion of calories in the Mediterranean diet that come from fat can reach 40 per cent or even higher (while carbohydrates typically account for less than 40 per cent).

Since the staple-nutrient fat has been demonised for so long and is still seen by many people as a kind of poison and the quickest way to gain weight, I think it's worth taking a closer look at the healthy, high-fat Mediterranean diet. The results produced in recent years by many scientific studies of this way of eating show that there is simply no rational basis for our fatphobia. To be clear from the outset: just as it is possible to live very healthily to a very old age on a high-carb diet, it is also possible to do so on a high-fat diet. The following is a graphic summary of main points so far:[9]

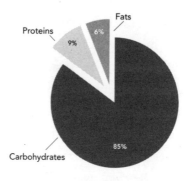

Traditional Okinawan diet

The traditional diet in Okinawa is extremely rich in carbohydrates, extremely poor in fats, and very healthy.

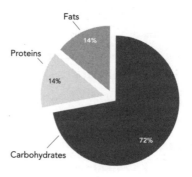

Tsimané diet

The diet of the Tsimané contains a slightly smaller proportion of carbohydrates and slightly more fat. It is also very healthy.

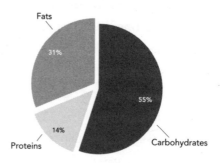

Adventists' diet

The diet of the Adventists is still carb-heavy. It contains relatively little fat, but still enough to mean it no longer warrants the designation 'low-fat diet'. It is also very healthy.

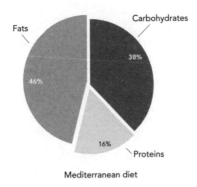

Mediterranean diet

The Mediterranean diet typically contains more fat than carbohydrates. And it is also very healthy.

The high-fat Mediterranean diet: so healthy they had to abandon the experiment early

The Mediterranean diet gets its name, of course, from the fact that it's based on the traditional eating habits of people who live in that part of Europe — in particular, in southern Italy and Greece, especially on the island of Crete. The name is somewhat misleading as there's no one, *single* Mediterranean way of eating. Just to be quite clear: someone visiting a branch of McDonald's looking out on the port of Genoa cannot claim to be eating 'Mediterranean food'. What I mean by this is that the modern eating habits of many Mediterraneans do not necessarily count as a Mediterranean diet in the sense that excites so many nutrition scientists.

Okay, that kind of goes without saying, but it actually goes a step further. Even your usual margherita pizza or spaghetti bolognese (which used to be one of my favourite meals) are *not* dishes that experts would describe as typical examples of food from a Mediterranean diet. Again: for nutrition researchers, pasta is *not* the be-all and end-all of the Mediterranean diet, despite the fact that we tend to equate Mediterranean food with huge piles of pasta. So you might argue that the term 'Mediterranean diet' is not particularly apt. Personally, I think life's too short to get upset about terminology. I think it's more important to be clear about what I mean when I talk about the Mediterranean diet in the context of modern nutrition science. So here are the key components of a Mediterranean diet:

- plenty of (local, seasonal) vegetables, pulses, and fruit
- a preference for wholegrain products, such as wholemeal bread
- regular portions of nuts and seeds
- wine with meals
- lots of extra-virgin olive oil
- little milk and few dairy products (mainly in the form of cheese and yoghurt, thus a preference for fermented dairy products)
- fish several times a week
- a preference for white meat such as poultry; red meat (pork, beef) only a few times a month
- up to seven eggs per week
- very few sweets and confectionaries (fruit is typically eaten for dessert)
- generous use of herbs and garlic as seasoning, and more sparing use of salt.[10]

You can easily work out your own 'Mediterranean factor' with the aid of a short questionnaire that's used in experiments to test how closely subjects adhere to the Mediterranean way of eating (see fig. 3.1). The more points you get, the more Mediterranean your diet is, in the idealised, nutrition-science sense of the term. The 'Mediterranean factor' resulting from this questionnaire is more than just a bit of fun. The higher your result, for example, the lower your risk of developing high blood pressure, diabetes, and obesity — and in particular, obesity in the abdominal region.

It's particularly remarkable that fats play such a significant part in this effect. We see that it is the very foodstuffs that are highest in fat content and which we usually consider fattening that are helpful in losing weight with the Mediterranean diet. Nuts, for example, have been associated in scientific studies with the *most reduced* risk of abdominal obesity. In other words, regularly snacking on nuts is more likely to help you achieve a flat tummy than not eating nuts at all. And according to such analyses, even olive oil is a 'slimming food'! (While, conversely, totally fat-free soft drinks are among the most fattening foods — more on that in the coming chapters.)[11] Results like these provide initial indications that the fat we eat is not automatically stored as fat by our bodies. Fat doesn't necessarily make you fat. And for some high-fat foods, such as nuts and olive oil, the opposite is true.

TEST YOUR 'MEDITERRANEAN FACTOR'

Question	If applicable, add 1 point
Do you use olive oil as your main source of fat when cooking?	Yes
How much olive oil do you eat per day?	At least 4 tablespoons
How many portions of vegetables do you eat per day? (1 portion = 200 grams)	At least 2 (of which at least one is raw vegetables or salad)
How much fruit do you eat per day?	At least 3 portions
How many portions of red or processed meat do you eat per day? (1 portion = 100 to 150 grams)	Fewer than 1
How many portions of butter, margarine, or cream do you eat per day? (1 portion = 12 grams)	Fewer than 1
How many soft drinks do you drink per day?	Fewer than 1

How much wine do you drink per week?	At least 7 glasses (of 100 mL, which is about one bottle)
How many portions of pulses (beans, lentils, chickpeas) do you eat per week? (1 portion = 150 grams)	At least 3
How many portions of fish do you eat per week? (1 portion = 150 grams)	At least 3
How many portions of sweets and confectionaries (cakes, biscuits, etc.) do you eat per week?	Fewer than 3
How many portions of nuts do you eat per week? (1 portion = 30 grams)	At least 3
Do you prefer white meat, such as chicken or turkey, to red meat, such as burgers and sausages?	Yes
How many times a week do you eat sofrito (a tomato, onion, garlic, and olive-oil sauce)?	At least 2

Fig. 3.1 This questionnaire is used by researchers to sound out how closely someone adheres to the idealised Mediterranean diet, which has been shown to be extremely healthy in many tests. The more points you scored, the more 'Mediterranean' your diet is. The risk of a serious cardiovascular event (stroke, heart attack) is more than 50 per cent lower for those who score 10 points or more, compared to those who score 7 or less. The three aspects of this diet that contribute most to lowering that risk are (in order of significance): vegetables, nuts, and wine.[12]

All of these results stem from purely observational studies, which do not take into account cause-and-effect relationships. For example, they can't give us information on whether nuts and olive oil are the actual *cause* of the beneficial effect recorded. However, there have now been several experiments that provide impressive confirmation of the observations made.

A few years ago, a team of Spanish researchers carried out a large-scale study on the effects of the Mediterranean diet with almost 7,500 test subjects. Half the subjects were instructed to follow a high-fat Mediterranean diet closely. The other half were the control group, and were called on to pursue a diet less rich in fat.

All the test subjects had an increased risk of a cardiovascular-disease event, and the researchers wanted to find out whether that risk could best be reduced by following a high-fat or a low-fat diet. 'What a question!' you might be saying right now, shaking your head in disbelief. Of course anyone who wants to protect their heart should avoid fat as much as possible.

To help the Mediterranean-diet group keep up their abundant consumption of fat, half of that group were given a free litre of olive oil every week to use as they wished. The other half were given free supplies of nuts (a 30 gram per day portion of mixed walnuts, hazelnuts, and almonds). In this way, the Mediterranean half of the subjects were subdivided into two groups — an olive-oil group and a nuts group. The low-fat half of the subjects were not given any free food supplies.

The results of the experiment were dramatic. It went so spectacularly well for the subjects who followed the Mediterranean diet and, by comparison, so badly for the control group, that the ethical review board recommended abandoning the experiment after a few years. In their opinion, the ongoing results showed it was no longer justifiable to continue to deny the control group a beneficial high-fat diet.

In particular, the Mediterranean diet drastically reduced the risk — relative to the control group — of suffering a stroke. It went down by 33 per cent in the olive-oil group and by 46 per cent for the group eating extra portions of nuts. Follow-up analyses also revealed that those who scored 8 or 9 points on the Mediterranean-factor questionnaire saw their risk of a severe cardiovascular event such as a stroke or a heart attack reduced by 28 per cent compared to those who scored 7 points or less. For those who scored between 10 and the maximum 14 points, that risk was reduced by no less than 53 per cent.[13] In short: the more Mediterranean your diet is, the better it is for your heart.

The Mediterranean diet had already gained a certain popularity thanks to earlier positive research results, but, when the Spanish study was published in the renowned medical journal *The New England Journal of Medicine* in 2013,[14] it became popular around the world (in fact, years earlier, a similar experiment in France had had to be discontinued early due to similarly spectacular results[15]).

And reports on the beneficial effects of the Mediterranean diet continue to appear regularly. Recent research, for example, shows a Mediterranean diet is associated with a measurable reduction in brain atrophy in old age.[16] For some people, such a diet has an astoundingly beneficial effect in combating depression.[17]

It's not only the Mediterranean diet itself that has risen in popularity; fat is also experiencing a real comeback thanks to these and many more positive research results. You might say low-fat diets are becoming increasingly passé, while fat has become increasingly in vogue in recent years. And understandably so! Mediterranean food is not only healthy, but also delicious, as I'm sure many people will agree.

This is due in no small part to the olive oil that features so predominantly in the Mediterranean diet, and not only because it tastes good itself. Olive oil also enhances the flavour of any other ingredients. Have you ever tried to clean oil or fat from a frying pan using only a kitchen sponge — without detergent? With water that's only lukewarm? Everybody knows fat is sticky; it smears and is hard to remove. And the same is true of its behaviour in your mouth. Fat gives food a pleasant consistency, and sticks flavours to your palate, as it were. It adheres to the inside of your mouth as it does to a pan, which allows it to fully develop its aroma rather than being washed down with the food immediately. So fat is a kind of natural flavour enhancer. Great!

Still: all this enthusiasm for fat should not belie the fact that even the mostly high-fat Mediterranean diet (there are also lower-fat versions) is 'only' one of the ways to eat more healthily. Incidentally, in the Spanish experiment, the proportion of fat in the diet of the Mediterranean group was 41 per cent, which may be high, but that figure for the control group, who had been asked to eat less fat, was still a relatively high 37 per cent. The fact that there was such a marked difference in the groups' risk of suffering a stroke shows once again that it is not the absolute amount of fat eaten that's important, but the overall character of the diet followed. In concrete terms: I don't believe that it is simply the high-fat nature of the Mediterranean diet that makes it so healthy — in the same way that, conversely, it is not the huge amount of carbohydrates in their traditional diets that's responsible for the excellent health and longevity of the Okinawans or the Tsimané. It's far more likely that the secret of healthy-eating cultures, from the islands of Okinawa to the rainforests of Bolivia and certain parts of the Mediterranean, lies in the fact that all those cultures eat a diet of real, natural food rather than industrially produced fast food. Their diet is sourced straight from nature and mainly, although not exclusively, from plants.[18]

Why it's so important that you shape your own diet

Most diet books side with a particular nutritional approach, supporting some 'program', whether it be vegetarianism, veganism, low-fat, low-carb, Paleo, pineapples, or some variation on the Mediterranean diet. They then proceed to 'prove', on the basis of some carefully selected studies, why their program is better than all the others.

On the one hand, such an extremely limited and partisan view can make life easier. But on the other, it means forcing people completely arbitrarily to follow *one* program, one doctrine of salvation. If you eat exactly like *this* (i.e. like Tsimané hunter, or an Okinawan islander, or like X or Y from some mountainous Mediterranean region), you will lose weight and live a long and healthy life. Yet there is no objective reason to follow any particular set of recommendations, since there are demonstrably many dietary approaches that can help you live long and stay healthy.

So we are justified in approaching this issue with a more open mind, and that strikes me as eminently sensible, since everyone's body is different. Rather than desperately forcing a diet on our bodies, we should be listening to them, feeling their responses, and experimenting with different ways of eating, irrespective of any dogmas and ideologies, to find out how our bodies react to various types of diet. This is the best way to find out what — within the bounds of the recommendable — is right for *you*. Slavishly following a specific diet usually only leads to abandoning it. There's no such thing as *the* ideal diet, even though most diet gurus and even 'official' institutions often insinuate that there is.

At first, the realisation that we are all different appears to make the task of finding a healthy diet more complicated. There's no one set of strict recommendations that are valid for everyone under all circumstances. At the same time, it means you have more options for shaping your own diet. Don't become a slave to any particular diet program by listening to some outside authority more than to your own body. Always bear in mind: your body is an authority, too.

What typically happens when we go on a diet? For a while, we try to keep it up, in defiance of the signals from our body — only to give up at some point, in frustration or disgust. No wonder! That diet is an alien program that we try to impose on our bodies. It seems to me that both the initial success and the prompt failure of many diets is inevitable due to their inflexible, pared-down nature. At the very beginning, they work because we're highly motivated — but also, and equally importantly, because the unusual fare we're eating simply goes against the grain. We don't really know what we should cook, the recipes aren't tasty, the food gives us wind, the food makes us feel sick … We eat less as a result and that of course leads to weight loss. But then sooner or later, we abandon the diet for *precisely the same reasons*. (The only time when this isn't the case is when we accidentally stumble upon a diet plan that happens to suit us perfectly.)

A good friend of mine tells me that every time she consumes too much olive oil and nuts (at my house), she has a sleepless night, plagued by a feeling that she is 'blocked up' inside. Clearly, a low-fat diet would be easier on her digestion. Forcing her onto a high-fat, low-carb diet would just be counterproductive.

As we know, the best criterion for a successful diet is that we can stay on it, which can only be the case in the long term if it doesn't feel too much like we have to 'endure' the diet. Thus, although it appears to complicate the matter, we can almost consider it a stroke of luck that there's no such thing as *one* route to diet success that is paved with gold, but rather many different pathways. After all, what it means is that you can largely put together your own diet plan, one that suits your own body. And that, in turn, makes it more likely that you will then stick to it.

As we have seen, there are limitations to the amount we can manipulate our protein intake. As far as carbohydrate and fat are concerned, there's more room to manoeuvre, at least in their proportions relative to each other, for the simple reason that those relative amounts are not important for most of us. What *is* more important is that we choose the healthy carbohydrates and fats. Recognising which carbohydrates and fats they are, and which we should be avoiding, is the subject of the following chapters. Let's start by looking at carbohydrates, and straightaway let's examine the most seductive, but also the most pernicious of the carbohydrates: sugar.

CHAPTER 4

Carbohydrates I: sugar — the two-faced carb

'Sugar scares me'
Lewis C. Cantley, world-leading US cancer researcher[1]

Sweet and sickly

Some of us know it from our own body's reactions, but everyone, without exception, who has children knows it from fraught personal experience: sugar is a very special substance. My four-year-old son, for example, is a highly sensitive sugar-detecting machine. He can sniff out sugar with a degree of sensitivity that would rival that of a Geiger counter calibrated to react to the tiniest amounts of radiation. I can — and this is no joke — make an astonishingly accurate estimate of the sugar content of the organic tomato sauce that's currently his favourite, simply from the level of passion with which he demands to have it at meal times. (If there are more than 10 grams of sugar per 100 grams of sauce, the Geiger counter hits maximum, and any attempt to persuade my son to accept a healthier alternative is totally doomed to failure, no matter what psychological tricks his despairing father tries on him ...)

This begs the question of what makes sugar so special. 'Oh, how sweet', 'sweet as sugar', 'sickly sweet', 'sugar addict' — just a handful of common phrases is all it takes to highlight our ambivalent attitude to this wondrous stuff. So what's so special about it? Let's take a closer look at the seductive, dangerous secret of sugar.

To start with, let me clear up any misunderstandings about what is meant by 'sugar'. The words 'sugar' and 'carbohydrate' are often used synonymously, which can lead to confusion. When I talk about sugar in this chapter, I'm referring to the crystalline white substance you buy at the supermarket in one-kilo bags to use in baking or to sweeten your tea and coffee. It's also sometimes called granulated or household sugar. The scientific name for this kind of sugar is 'sucrose'. In normal life, we usually just call it 'sugar'.

But household sugar is just one of many different 'sugars'. The technical term for the entire family of different kinds of sugar is the 'saccharides', and they are, indeed, more commonly referred to as carbohydrates. They come in all shapes and sizes. There are various 'monosaccharides' (also called 'simple sugars'), such as glucose and fructose, which I will examine in more detail soon. You can think of the monosaccharides as individual Lego bricks. In this metaphor, glucose could be a green Lego brick, and fructose could be a yellow one. Monosaccharides and other kinds of sugar that are made up of very few building bricks are sometimes referred to as 'simple carbohydrates'. Then there are complex sugars, also known as 'complex carbohydrates' or 'polysaccharides', which are made up of many building blocks. The prime example of a polysaccharide is starch, which is made up of thousands of glucose molecules linked together in long chains — so, long chains of green Lego bricks attached together.

Let's focus for now on household sugar, sucrose, which I will simply refer to by its normal, everyday name, 'sugar', for the rest of this chapter. Many people are surprised to find out that even this 'normal' sugar is made up of not *one*, but two different sugar building blocks: glucose and fructose. One green Lego brick attached to one yellow one. That means every sugar molecule is — to mix metaphors — two-faced, and that's where much of the hazard lies.

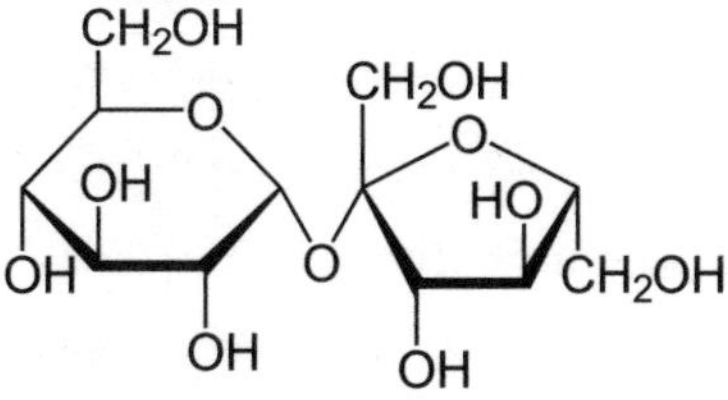

Fig. 4.1 Normal household sugar is a double sugar (disaccharide), which means it's made up of two simple sugar molecules (two monosaccharides): one glucose molecule (left) and one fructose molecule (right). The two molecules are very similar but not identical. As you can see in the diagram, glucose is hexagonal in shape, while fructose is pentagonal. The molecules are linked by the O in the middle, which represents an oxygen atom. H stands for hydrogen atoms, C for carbon. There should also be a C at each intersection of the lines (they are left out of the diagram for the sake of simplicity). Household sugar is just one of many different 'sugars', known technically as 'carbohydrates'. Roughly speaking, carbohydrates can be described as having a carbon-to-water-molecule (H_2O) ratio of 1 to 1. Or, written as a chemical formula: $C_x(H_2O)_x$. If, for example, X equals 6, then the resulting molecular formula will be $C_6H_{12}O_6$. Thus, as their name implies, carbohydrates are 'hydrated' ('watered') carbon atoms.

Glucose is sometimes called 'grape sugar', and fructose is also referred to as 'fruit sugar'. This term is also somewhat confusing since fruit always contains both glucose and fructose. Yes, grapes contain glucose, but they also contain a more-or-less equal amount of fructose (about 7 grams of each in every 100 grams of grapes). And by the same token, glucose is found not only in grapes, but in all fruit as well as in vegetables and, in fact, in all plants. Glucose is the basic substance that makes up such un-grape-like foods as bread, pasta, rice, and potatoes. To avoid any confusion, I will stick with the words 'glucose' and 'fructose' when referring to these sugars.

Okay, just one more time — perhaps because the name 'fruit sugar' reminds us of healthy fruit, many people are completely astounded to learn that it's the fructose part of sugar that's capable of damaging our bodies, and it does so in a unique way. As the saying goes, the dose makes the poison, and that adage is certainly apposite here.

And it's all the more apt since the speed with which sugar enters our system is such an important aspect. The faster the sugar molecules are transported into our bodies and reach our liver, for example, the worse the consequences are. In this respect, a comparison with alcohol is realistic: we all know that downing half a bottle of wine in one go on an empty stomach has a different effect to enjoying exactly the same amount of wine over an entire evening along with a five-course meal.

There is another similarity between sugar and alcohol. Like alcohol, sugar reduces our stress responses and calms us; it can have a comforting and even euphoric effect on us, which explains why we reach for the chocolate and ice cream when we are frustrated or lovesick. Some people tend to eat more when they're under stress, others eat less, but remarkably — as borne out by scientific experiments — *everyone* has a preference for sugary food when they're stressed![2]

While our brain succumbs to the sugar rush, our livers suffer in silence. Glugging down half a litre of cola or fruit juice to quench our thirst has the same effect on our liver as if we'd downed half a litre of wine. The reason for this can be summed up in one word: fructose.

Fattening fructose, or: preparing for a winter that never comes

Let's trace the route taken by a drink of cola, sweetened iced tea, lemonade, or fruit juice from the mouth to the inner body. This is the journey it takes: since it's a liquid, it doesn't need physically breaking down further; so after a short stay in the stomach, it rushes on into the small intestine, where the two parts of sugar are separated (unless, that is, the glucose and fructose are already separated in the food itself, as is sometimes the case with fruit or industrially produced high-fructose corn syrup (HFCS), also called 'isoglucose' or 'glucose-fructose syrup'). The connected glucose-fructose molecules become molecules of pure glucose or fructose, and the sugar starts to act in its two-faced way.

The separated molecules are small enough to pass through the gut wall and into the hepatic portal vein, which transports them to the liver. There, in the liver, the two monosaccharides part ways. While glucose puts on one face and leads a pretty conventional existence, fructose, by contrast, puts on another face and is far more extravagant in its behaviour.

When the liver requires energy, it treats itself to some of the glucose. As soon as the liver is 'full', it leaves most of the glucose molecules alone and lets them go their own way. The glucose then spreads through the rest of the body via the bloodstream, where it can be taken up by any cells that happen to need some energy at that time — from muscle cells to the cells of the brain. The brain loves glucose and gobbles up most of it. So far, so normal, as biological processes go. Many of the most notorious carb-bombs — bread, pasta, rice, and potatoes — contain little or no fructose. They contain mainly glucose in the form of starch. Glucose can be used as an energy resource by any of the body's cells.

However, sugar, honey, soft drinks, fruit, and fruit juice contain not only glucose, but also fructose — which accounts for roughly half of their sugar content. That fructose is treated quite differently by the body. Fructose molecules are also transported to the liver by the portal vein, but, once they arrive, something very strange happens to them. No matter how full the liver is, it absorbs almost the entire flood of fructose like a sponge, transforming some of it into fat inside its cells. This means that, for our liver, fructose is not at all the same as glucose, despite the fact that they have the same calorific value — which means that they release the same amount of energy when burned in a steel container (this is the way scientists measure calorific value — one calorie is defined as simply the amount of energy required to heat one kilogram of water by one degree Celsius).

Nobody knows why our bodies process fructose in this particular way. It must have something to do with our evolutionary history — indeed, it's probable that this mechanism was once lifesaving for us. The molecular biologist Lewis Cantley came up with a speculative explanation that seems plausible to me. Cantley is one of the leading cancer researchers in the US. He discovered a protein molecule that is part of the insulin and mTOR signalling pathway, a discovery that earned him the Breakthrough Prize in Life Sciences in 2013, which comes with a purse of three million dollars (far more than the Nobel Prize). The award is funded by, among others, the co-founder of Facebook Mark Zuckerberg and the co-founder of Google Sergey Brin. Cantley says:

> Fruits ripen just at the end of the growing season, which generally means, in almost all environments, that you're not going to have much to eat over the next few months. So the best way to survive is to convert everything you eat at that time into fat … That's why fructose was spectacular for us 10,000 years ago, getting us through these famines that we faced every year. But today we don't have famines and so we just get fat.[3]

Animals usually regulate their body weight with extreme rigour. An animal's body 'wants' to be neither too thin nor too fat — unless the animal is preparing for a long period of scarcity, or, in the most radical case, a period of hibernation. At such a time, the animal will try to store as much energy as possible — precious energy, which its body stores in the form of fat reserves.

According to one hypothesis, when an animal or a human being ingests more than a certain critical amount of fructose, the fructose is not simply transformed into fat. No, the glut of fructose acts like a kind of warning bell, telling the body that winter is coming. So what does the body do when it hears that alarm ringing? It switches to ultimate storage mode: it now prefers to put everything we eat aside for leaner times, in the form of fat. Thus, in a way, fructose turns on a kind of 'fat switch' inside us, activating a prehistoric energy-saving program.[4] This is what happens to us every time we drink a lot of cola or fruit juice.

We could well continue to speculate in this direction: it would be a clever evolutionary strategy to make us love the taste of fructose so much that we become properly greedy for it. Since the fruit trees and bushes will be eaten bare at some point, we would soon no longer be able to satisfy that greed or craving. Although we would have to deal with a bad case of sugar withdrawal when winter eventually came, we would at least have put on a warming layer of fat for the cold, lean season. Of course, this is all pure speculation.

If there is any grain of truth to these speculations, we could say that this strategy employed by our bodies for getting through the winter — triggered by orgies of gorging on autumn fruits — helped our Stone Age ancestors gain an advantage in the battle for survival. In our modern world, where sugar is everywhere (it takes a real effort *not* to eat it), that useful strategy becomes unhinged and turns against us. Thanks to a food industry that tends to spice up its offerings by stuffing them with as much sugar as possible, our bodies constantly think we are facing an imminent, Siberian-style Ice Age. All year round, our bodies are preparing for a calorific winter that never comes.

When I questioned the researcher Lewis Cantley more closely about this, he said he believes that the food industry makes deliberate use of our evolutionarily preprogramed 'addiction' to sugar. 'The worst part is that the food industry has every incentive to add sugar to everything since it is the cheapest ingredient,' says Cantley. 'It takes advantage of the addiction, thereby increasing sales.'[5]

The strange way our bodies metabolise sugar casts further light on how this addiction to sweet things might be driven. An immediate consequence of the fact that fructose — one half of sugar — is processed almost exclusively by the liver is that half the energy contained in extremely energy-rich 'foods' such as cola and sweets never reaches the command centre we call the brain. It is intercepted by the liver and turned into fat. No wonder the brain reacts by sending out the message that we should drink another glass or eat another snack, because the brain still needs more glucose. To satisfy and satiate the brain, it takes *twice* the amount of sugar compared to when we eat regular starch or pure glucose.[6]

Many of us are all-too familiar with this phenomenon: a soft drink, a glass of juice, or a pack of gummy bears doesn't really fill you up, despite the high calorie content of such delicacies. Of course, we do eventually stop snacking or drinking, but, as the US science writer and sugar critic Gary Taubes puts it, we stop either when we feel guilty or when we start to feel physically sick.[7]

It's no accident that not a single dietary approach among those that can be described as credible advises eating unrestricted amounts of sugar. On the contrary, whether it's the low-carb, low-fat, Mediterranean, or Paleo diet — in all of them, sugar is the first to bite the dust. Every healthy eating program pleads for sugar to be cut, to a greater or lesser degree. I find the scientific studies that reveal the negative effects of sugar to be very convincing. I follow the basic principle — the less sugar eaten, the better.

The reason for this is that, even should the fat-switch hypothesis turn out to be wrong, the fact remains that sugar does not provide our body with any nutrients along with the large portion of calories it contains. These are often described as 'empty calories', which is a strange expression. It insinuates the idea that sugar is somehow a source of pure energy. From this, it appears to follow that the negative effects of sugar stem only from the fact that it displaces other, more nutritious food in our diets, since we all need a certain amount of nutrition and energy.

However, in view of sugar's adverse metabolic effects, the word 'empty' is clearly an understatement and far from being an apt description of the real effects sugar has on our bodies. As a comparison: alcohol is also rich in energy and can also displace more nutritious foodstuffs from our diet, which often happens in the case of alcoholism. Few people who have ever visited Munich's Oktoberfest, however, would entertain the strange idea that the calories contained in alcohol might best be described as 'empty'. Far more importantly, alcohol is extremely harmful in high doses, irrespective of whether it displaces other, healthier foods from a drinker's diet. What I mean to say is that the calories in sugar are not 'neutral'. They are not only harmful because they stop us eating other nutrients; they are harmful in themselves. Dietary approaches such as low-fat or Paleo can probably be considered better than our normal way of eating for the simple reason that they include very little sugar (the same is true, of course, of the traditional Okinawan diet[8]).

What does this mean in our day-to-day lives? Ultimately, it's all about the dose. The way I see it, the occasional piece of cake, a scoop of ice cream on a hot summer's day, or a sweet dessert is probably not going to be a problem, especially if it comes in the form of the irresistible crème brûlée my wife sometimes — but all too rarely — makes for me. Even a spoonful of sugar in your coffee or tea is not going have that much of an impact when you bear in mind that half a litre of cola contains 14 such spoonfuls (one teaspoon of sugar corresponds to about four grams). And you might not believe it, but the healthy-looking, naturally cloudy organic apple juice that I have in my fridge at the moment (I'm not sure who had the audacity to smuggle it into my home) contains just as much sugar as cola! So, in practical terms, that means: beware of sugary, sweetened drinks. The biggest source of sugar in our diets is soft drinks such as cola and lemonade, as well as the many popular 'energy drinks', and even 100 per cent fruit juices. They all contain disproportionately large amounts of sugar, and the fact that they come in liquid form means the sugar they contain also enters the bloodstream almost as fast as if it were administered by intravenous infusion.

Personally, I avoid all such sugar infusions. Every now and then, I drink a glass of freshly squeezed orange juice, or a glass of pomegranate or beetroot juice. Although I regularly use honey (which is also a mixture of fructose and glucose), I use it sparingly and usually only in salad dressing. Unfortunately, the frequently praised alternatives, such as the agave syrup so popular with vegans, are no better — agave syrup is made up almost exclusively of fructose, which is why it's so unbelievably sweet: fructose has a much sweeter taste than glucose.

So the only remaining options for sweetening up our lives are artificial sweeteners such as aspartame, saccharin, and sucralose. How should we rate them? Are they good alternatives to sugar? Firstly, in general, these sweeteners are broadly judged to be 'safe'. One argument put forward to support this judgement is that, as synthetically produced substances, they can't be metabolised by our body. On the whole, that's true, but it overlooks one little detail: 'we' are not made up only of 'us'.

Deep in our guts, numerically most common in the large intestine, billions and billions of bacteria silently live out their little lives. This microbial population inside us is called our 'microbiome',[9] and makes up one to two kilos of our overall body weight. And just like us, it gets hungry. Very hungry. But these poor bacteria are at the back of the queue, as it were, and only get to eat the indigestible leftovers from our small intestine. Those leftovers can be very healthy substances, such as dietary fibre from wholemeal bread, but they can also include substances that are not so healthy.

According to preliminary research results, artificial sweeteners are among those substances. For example, an Israeli team of scientists based at the Weizmann Institute of Science demonstrated that consuming artificial sweeteners can massively disturb the balance between different bacteria strains in the gut in just a few days. Less beneficial strains spread, while beneficial ones (such as *Lactobacillus reuteri*) recede. The negative consequences of this disturbance in the microbiome balance in our gut reach far beyond our large intestine, to our entire body. The most important effect of this imbalance is to reduce our ability to process the glucose in our blood, which is the first step on the road towards diabetes. Furthermore, the indications are that artificial sweeteners can contribute to obesity (there are contradictory scientific findings around this question). Thus, ironically, replacing sugar with artificial sweeteners to ward off obesity and diabetes may have the opposite effect. This is reflected in the summary published by the Israeli researchers in the science magazine *Nature*:

> Artificial sweeteners were extensively introduced into our diets
> with the intention of reducing caloric intake and normalizing

blood glucose levels without compromising the human 'sweet tooth' … This increase in NAS [non-caloric artificial sweetener] consumption coincides with the dramatic increase in the obesity and diabetes epidemics. Our findings suggest that NAS may have directly contributed to enhancing the exact epidemic that they themselves were intended to fight.[10]

In short, artificial sweeteners are really not to be recommended! You may now still be asking whether it's necessary to deny yourself all sugar and sweeteners with religious zeal. One answer is, no, probably not. But another answer is that even if you cut down on using sugar directly yourself, you will still almost inevitably be eating a lot of it because sugar — as Cantley says — is in almost everything we buy to eat. It's in bread, yoghurt, sausages, ham, ketchup, organic pasta sauce, cornflakes, muesli, and much, much more. Artificial sweeteners, meanwhile, can be found in all those 'diet' and 'light' products. I don't think you should stop eating individual products that are otherwise very healthy, just because they contain a couple of grams of sugar, but whatever you choose, it makes sense to be very aware of the scale of sugar use in the food industry.

This is another important aspect, since stirring a couple of cubes of sugar into your coffee or tea *yourself* or using sugar when *you* bake your favourite lemon-drizzle cake is something that you can at least be consciously aware of, and you can directly monitor the amount of sugar you use. You can choose to use a little less sugar if you want. But industrially produced food mostly smuggles its sugary contraband into our bodies, and it does so with practically every meal we eat. There's nothing wrong with a sweet dessert after dinner, but if *everything* is covertly turned into a dessert, then something has clearly got out of hand. Take a closer look next time you are out grocery shopping. This is another reason why it's such a good idea to cook for yourself as far as possible, using fresh ingredients plucked straight from nature rather than from a factory.

Incidentally, I have found a replacement for sugar in my coffee and tea that's not only a healthier alternative, but also a tastier one, in my opinion. I now always have a generous piece of very dark chocolate with my coffee. Such chocolate contains relatively little sugar — 7 grams, so the equivalent of 'only' two spoons of sugar, in an entire 100-gram bar of my preferred 90 per cent dark chocolate. Milk chocolate contains about 50 grams of sugar, which is seven times as much. Dark chocolate in particular contains bioactive plant substances called 'flavonoids', which improve blood-vessel function, reduce blood pressure, and increase insulin sensitivity (more on this soon).[11] Alongside nuts, dark chocolate is my absolute favourite snack!

When I drink tea, I often have some fruit with it (my tip: a cup of Sencha Uchiyama[12] is delicious with an apple). Fruit? Doesn't that also contain fructose? Yes, it does, but in reasonable amounts. And anyway, I've never met anyone who would willingly wolf down five or six apples in succession, or eat two kilos of grapes in one sitting (although some hungry Stone Age humans may have done precisely that as winter loomed). Squeezed into the form of apple or grape juice, it's easy to consume that much in a few seconds. Eating three or four portions of fruit a day should not be a concern as far as the sugar content is concerned. Personally, my favourite kind of fruit is berries, such as blueberries, strawberries, raspberries, and blackberries, which don't contain much sugar but are full of beneficial substances that, among other things, inhibit the absorption of sugar by the small intestine, thereby helping to prevent a sugar shock.[13] So there's no need to be wary of whole fruits.

Another point about fruit is that the sugar it contains comes in an integral structure together with dietary fibre. The sugar molecules are only gradually liberated from the fruit and released into the bloodstream. That means the liver isn't overwhelmed by a sugar tsunami. When the fruits are squeezed for juice, on the other hand, most of the fibre and other beneficial substances are lost. What's left is mainly water — and sugar. Smoothies are a little better in this respect, but such thorough liquidising also destroys the structure of the fruit to such an extent that there's little to prevent the sugar from being digested rapidly, causing highly concentrated sugar to enter our bodies quickly. Even if the nutrients as such stay in the juice or smoothie, a whole piece of fruit is more than just the sum of the nutrients it contains.[14] I used to consume fruit juice and smoothies in copious amounts, not least of all because I believed they were good for my health. These days I (almost) never drink my fruit — I eat it.

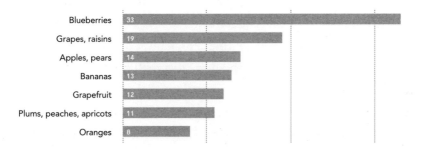

Reduced risk of type-2 diabetes by percentage

Fig. 4.2 Eating fruit — in the form of whole fruits — can lower the risk of developing type-2 diabetes. Fruit juice, on the other hand, is associated with an increased risk. Here we see what happens when three glasses of fruit juice per week are replaced with a comparable amount of whole fruit. Replacing the juice with blueberries, for example, causes a reduction in the risk of diabetes of over 30 per cent. These calculations are based on the dietary data of more than 150,000 women and 36,000 men.[15]

From fatty liver disease to cell ageing

Two-faced sugar is not only fattening, it also — in large amounts over a long time — makes you sick, although the two effects often go hand in hand and can't easily be separated from one another. As outlined earlier, when the liver is flooded with fructose it turns that fructose into fat. Worse — the more fructose we eat, the more practised the liver becomes in digesting the stuff. It begins to adjust to the situation, by activating certain genes, perfecting its ability to turn fructose into fat.

However, like all the other organs in our body, the liver is not designed to store fat — that's the job of the adipose tissue just beneath our skin. Fat that accumulates excessively in places it shouldn't be is aptly known as 'ectopic fat' (from the Greek *ektos* 'out' and *topos* 'place', i.e. fat that is out of place. Intra-abdominal fat, which accumulates *around* our inner organs, is a kind of ectopic fat in the broadest sense, but the term is usually used in its narrower sense to describe fat deposits *inside* the cells of the liver, pancreas, muscles, and so on.) Ectopic fat is harmful because it prevents the cell it has occupied from performing the cell's intended purpose.

The most serious malfunction as far as the consequences are concerned is one which I will return to again and again, mainly because it's so common. I'm talking about insulin resistance. When fat accumulates in the cells of the liver, they become less sensitive to the hormone insulin. The insulin signalling pathway within the cell is thwarted. That means the pancreas is forced to secrete more of the hormone to compensate for this reduced sensitivity. In this way, insulin resistance leads to increased levels of insulin in the blood. As I mentioned before, insulin is a fat-storage hormone. That means, firstly, we are well on the way to putting on weight (especially if, as is usually the case, our cells' sensitivity to insulin has *not* yet been compromised).[16] And, secondly, diabetes is looming on the horizon.

Thirdly, cancer researchers like Lewis Cantley are convinced that sugar can even increase the risk of developing the disease via the insulin pathway. This is because insulin is also a hormone that stimulates growth (similarly to mTOR, and in fact directly connected to it: insulin activates mTOR via the very protein molecule that Cantley discovered). Fatally, our normal cells are not the only ones to be stimulated by insulin into engaging in all sorts of building activities — most kinds of cancer cell also have a kind of reception antennae that are tuned to insulin and react to the hormone by growing and multiplying. As Cantley explained to me about insulin, 'in addition to telling the tumor to take up glucose, it also tells the tumor to take up amino acids and to synthesize amino acids and nucleic acids and proteins and fats, all the things that a tumor needs to grow!'[17]

Incidentally, as a cancer researcher, Cantley himself avoids sugar wherever he can for this reason.

Let me summarise the main points again: regular floods of fructose cause fat to accumulate in the liver. The fat in its cells makes the liver less sensitive to the hormone insulin, which leads to an increase in insulin secretion, which in turn abets the development of many different medical conditions — from obesity to cancer.[18]

Again, the liver is not designed for stockpiling fat. On the contrary, the liver provides the body with energy in the form of glucose and fatty acids. Thus, the liver wants to distribute the extra fat throughout the body, which, in principle, could well use the energy it provides.

The liver proceeds with the distribution in an orderly manner and loads the fat onto specially created transport molecules. You might picture those transport molecules as tiny little buoys, floating through the bloodstream with their cargo of fat. (The buoys are also loaded with cholesterol, a substance that is very similar to fat. Cholesterol is also required by our cells because, among other things, it's a component of cell walls.)

Why does the liver go to the trouble of assembling costly transport buoys? Because it would not be a good thing if our blood were to turn into something resembling chicken soup. It's a well-known fact that fat and water don't mix well, so any free-swimming fat molecules would form clumps in our water-based blood like the blobs of grease on the surface of a pot of soup. To stop that from happening, the liver loads the fat molecules onto those little buoys. When the transportation buoys are loaded with their cargo of fat molecules (and cholesterol), the liver launches them into the bloodstream. They then embark on a lengthy journey through the body, knocking on the doors of various organs to enquire whether they require any fat right now.

Muscle cells will typically say, 'Ah, energy, great, give it to me!' And they indulge in a bit of the fat. However, with the kind of oversupply of fat we're talking about here, the muscle cells often take on more fat than they need. This results in the fat accumulating in the muscle cells as well, making them also insensitive to insulin. The diagnosis of diabetes is getting closer and closer.

The transport buoys made by the liver and sent out into the bloodstream are known medically as VLDL particles (very low-density lipoprotein particles, which can be described as a protein frame carrying various different fats and having a very low overall density). A VLDL particle acts like a delivery truck, transporting fat molecules and cholesterol to our organs and the cells of our body. Since VLDL particles give up their cargo of fat as they travel through the body, they become successively smaller, until they eventually turn into LDL (low-density lipoprotein).

However, the liver also synthesises transport vehicles especially for the purpose of collecting surplus cholesterol from the cells of the body and taking them back to the liver. This kind of transport molecule is known as HDL (high-density lipoprotein). As a general rule, we can say that high levels of LDL in the blood are bad, while high levels of HDL are good. A high-sugar diet raises the level of LDL in the blood and lowers the level of HDL cholesterol. The many LDL particles along with their cholesterol accumulate in the walls of our arteries, causing inflammation and eventually leading to blockages. Put simply: it's not just fat in our diet that blocks arteries, but also sugar.

What's particularly fatal is that when the liver is rapidly flooded with fructose (and by extension this is true of all rapidly digested carbohydrates), it reacts as we might expect, by synthesising relatively large numbers of fat particles, which are known medically as 'triglycerides'. To rid itself of this glut of fat, the liver loads as many triglycerides onto its VLDL transport vehicles as possible. This creates — and this is the crucial point — extremely large, literally very fat VLDL buoys.

As they pass through our body, these super-fat VLDL cargo buoys, like all VLDLs, unload their triglycerides into our cells, getting ever smaller as they do so, as if they were on a diet themselves. Although they started off very fat, these large VLDL particles in particular become very small LDL particles after losing their fat, which means they can then filter into the walls of our arteries like fine sand. It's those tiny LDL particles (medical term: small, dense LDL or 'sdLDL') that are increasingly being revealed to be extremely harmful. It has become increasingly clear in recent years that they increase the risk of suffering a heart attack to an extremely high degree — more than other, larger LDL particles. In this way, sugar plays a significant part in clogging our arteries and, consequently, in increasing the risk of a heart attack.

It's actually shocking to see how much consuming a lot of sugar increases the risk of dying of a heart attack. One study in which Harvard University participated found that people who obtained between 10 and 15 per cent of their calories from food that contained added sugar — usually in the form of soft drinks, sweet desserts, fruit drinks, and candy — had a 30 per cent higher chance of dying from cardiovascular disease. The mortality risk of those who obtained 25 per cent or more of their calories from such foodstuffs was increased almost threefold![19]

Time and again, soft drinks stand out as particularly bad in such studies. It really is no exaggeration to say that cola, lemonade, and other fizzy drinks are not only the least healthy source of carbohydrates, but among the least healthy 'foods' overall (as far as 100 per cent fruit juices are concerned, the results are much more mixed and more disputed, probably because they at least still contain the bioactive substances in the fruit they were squeezed from, which counter the harmful effects of the sugar in them).

Just as importantly, soft drinks probably also drive the ageing process. A recent analysis, which included the involvement of the Australian-American Nobel laureate Elizabeth Blackburn, found that the more soft drinks a person consumes, the shorter their telomeres are.

Anyone who has heard of telomeres knows that shortened ones are bad news. The genetic material in our cells is usually contained in our chromosomes, a kind of biochemical package. The ends of our chromosomes are pretty sensitive fellows, but, luckily, they're protected by our telomeres — which act a little like the plastic or metal aglets that stop our shoelaces from fraying. Every time a cell divides, those protective sheaths at the end of the chromosomes get a little shorter. In simple terms: the shorter the telomeres, the older the cell. At some point, the telomeres become so 'worn out' that the cell dies.

Blackburn and her colleagues discovered that drinking one glass of cola or lemonade or similar — not even a quarter of a litre — per day ages a person's cells (as measured by the length of their telomeres) by 1.9 years. Consuming just over half a litre of fizzy drinks can lead to an additional 4.6 years of cell ageing. That's equivalent to an increase in ageing as measured by telomere length that's usually only associated with smoking![20]

Test: what does drinking one litre of cola per day do to our body?

Many of the metabolic processes I've described were discovered over the decades using animal models.[21] However, many of the findings about excessive sugar consumption and the harm it can do to us humans are based on observational studies, which should always be viewed with a certain scepticism. It's quite possible that people who love cola and similar drinks may not exactly be the biggest fitness freaks on the planet, while most health-conscious people shun soft drinks. In that case, how can we be sure that it really is the sugar that leads to accelerated ageing and premature death, rather than their generally unhealthy lifestyle? This brings us back to the old question of causation vs correlation.

A further complicating factor is that sugar is also high in calories and so eating a lot of it also contributes to obesity, and the resulting obesity could also be responsible for some of the problems observed. Coca-Cola and soft-drink makers in general love this argument. After all, it means that the problem isn't that *they* manufacture products that might be just as harmful as cigarettes, but that *we* are ultimately at fault. We are simply unable to control ourselves — and we don't exercise enough! In other words, we're both insatiable *and* lazy.

In view of all this, clarity can only be gained from tests on human subjects, although such trials almost verge on bodily harm. Still, it's only with the help of such experiments that scientists can test their metabolic models and find out whether sugar — consumed in daily quantities that aren't completely outlandish — really is the *cause* of the health problems observed. And it's only in recent years that a few such — complicated and expensive — experiments have been carried out. And by and large, they confirm our fears. In the following, I will highlight two trials that I consider particularly relevant, and which appear to be particularly rigorous and convincing.

In the first experiment, Danish researchers randomly divided just under 50 overweight test subjects aged between 20 and 50 into four groups. All the participants were asked to continue their normal eating habits for the next six months, but with one small adjustment: the first group were to drink a litre of cola per day, the second group were asked to drink a litre of low-fat milk (1.5 per cent fat) every day, the third group were tasked with drinking a litre of diet cola, and the fourth group a litre of water.

Of course, the comparison with diet cola and water wasn't a fair one, since they contain no calories. But that made the question of whether there would be a difference between the first and second groups (cola vs milk) all the more interesting, since cola and low-fat milk contain an almost identical number of calories (the cola used in the experiment had 440 calories per litre; the milk had slightly more, at 460 calories).

And indeed, after six months, there were marked differences between the four groups. The blood-triglyceride levels of the cola drinkers had risen by 32 per cent — as predicted by metabolic models — and their overall cholesterol levels went up by 11 per cent, while neither of those metrics changed in the other three groups. This difference was even more pronounced when the participants' livers were examined: the accumulation of fat in the livers of the cola drinkers was up by 143 per cent compared to the milk drinkers. The overall gain in fat mass was similar in all four groups, but the daily dose of cola had caused those subjects to gain intra-abdominal fat. When the VLDL transport buoys with their triglyceride cargo go knocking on doors throughout the body to distribute their fat, they apparently prefer to unload their fatty freight in places where it doesn't belong and where it can do harm.[22]

In summary, the findings gathered support the following cascade:

1. Continuous consumption of large amounts of sugar leads to a fatty liver. This makes the liver insensitive to the hormone insulin.

2. The liver attempts to rid itself of the surplus fat by transporting the fat molecules with the aid of transport particles (VLDL) heavily loaded with triglycerides around the body, including to the muscles, which then also become fatty and insulin resistant.

3. The pancreas increases its production of insulin as a response to insulin resistance. It pumps more insulin into the bloodstream, raising the risk of developing obesity, cancer, and other problems.

4. The surplus liver fat is also deposited in the abdominal region. This abdominal fat becomes inflamed, and the resultant, harmful inflammatory substances further boost insulin resistance and diseases associated with ageing.

5. As they pass through the body, the extremely fat VLDL particles turn into small, dense LDL particles (sdLDL), some of which accumulate in the artery walls, blocking them and once again raising the risk of a heart attack.

So we're confronted with a whole complex of medical problems, the core of which is formed by fattening of the liver, insulin resistance/ diabetes, obesity, and an extremely elevated risk of cardiovascular disease.

Other experiments suggest that this devastating metabolic breakdown is principally due to the fructose part of sugar. Those include a ten-week study in which the test subjects were divided into two groups, one which drank three glasses of a high-glucose drink per day, and another which drank three glasses of a high-fructose drink. All the participants had gained weight by the end of the ten-week period, but a closer inspection revealed dramatic differences between the two groups of subjects. In the glucose group, the surplus calories were stored directly under their skin — where fat is supposed to be stored. In the fructose groups, most of the extra calories were stored as intra-abdominal fat. Furthermore, the fructose had led to increased fat accumulation in the liver and insulin resistance, and the longer the 'fructose diet' continued, the higher the subjects' dangerous sdLDL levels climbed (all this was not observed in the glucose group).[23]

Sugar: conclusion

What conclusions can we draw from all of this? Perhaps this, although only cautiously: in the past it has proven counterproductive to pick out one food group, such as fat, and blame it for everything. For the longest time, the promise was that if you could just stay away from fat, especially saturated fat, then everything would be fine. And? What did we do? We listened to the promise and, among other things, tried to eat more fat-free products, which are often full of added sugar! With hindsight (when, as we know, it's easy to have 20/20 vision), this completely well-meant advice has turned out to be very short-sighted. One not-insignificant side effect of this fatphobia is a fatal increase in the acceptance of sugar and other processed carbohydrates. Of course, that was not the intention, but, when it comes down to it, we have to eat *something*. So we need to beware of this demonisation trap.

Still, it seems to me that the warnings about eating too much sugar are slightly different in nature. Firstly, it's not about condemning an entire food group, but a very specific substance. And it's not even that sugar or fructose is 'evil' per se, but rather it's that the extremely unnatural, historically unprecedented amounts we consume today damage our bodies over the long term. And the breakneck speed with which we sometimes shovel these large amounts of sugar (via cola, fruit juice, etc.) into our bodies makes a not-insignificant contribution to their harmful effects.

My conclusion is that steering clear of soft drinks and exercising restraint with sweets and fruit juice is probably half the battle when trying to eat more healthily. If you also take care that your breakfast is not really a dessert and your dinner is not actually a pudding in disguise, you can take all the more pleasure in eating something sweet after your meal.

CHAPTER 5

Carbohydrates II: why some people's bodies only react well to a low-carb diet

The man who tried to eat himself to death — and failed with great success

In his earlier life, Sten Sture Skaldeman was fat, and he just kept getting fatter and fatter. In his attempts to lose weight, he stuck to the 'official' dietary advice. He never ate butter. He ate lots of bread, pasta, and polenta. And the result was: he gained more weight. By the age of 40, Skaldeman weighed more than 100 kilograms. In his desperation, he tried starvation diets — which always resulted in short-term weight loss, before his weight rebounded and eventually exceeded 125 kilos.

Skaldeman started studying diet books. In one, he read: you won't get fat if you don't eat fat. Skaldeman followed that 'principle', but still gained weight. By then, the 1.75-metre tall Swede weighed in at 150 kilos. The pain in his joints became unbearable, and Skaldeman sometimes required help just to put his shirt on. His heart was also in a bad way. His overworked ticker was barely able to pump blood around his massive body. Even walking to the letterbox was a challenge;

Skaldeman had to stop to catch his breath every few metres.

Visiting a clinic, Skaldeman was told he had the highest blood pressure ever measured in that hospital.[1] A cardiologist diagnosed heart failure. And, particularly informative for the focus of this chapter, the doctors also found he had astronomically high fasting blood-insulin levels. In other words, the cells of Skaldeman's body were so full of fat that they had become numb to the effects of insulin, which caused his pancreas to try to compensate by pumping out extremely large amounts of the hormone. 'Life was hell,' is the short but dramatic way Skaldeman described his situation to me.[2]

Finally, one day, after losing yet another weight-loss wager, he decided he'd had enough. That was in autumn 1999. Skaldeman was not yet 60, and was the father of four children. Despite that, he was at the end of his tether and decided to ignore all dietary advice from then on. He would just let himself go. He would simply eat whatever he wanted. And he accepted the fact that he would probably eat himself into an early grave.

And so Skaldeman changed his diet overnight. No more rules, no more prohibitions! Now he had bacon and eggs for breakfast, enjoyed his favourite meat for lunch, and for dinner, just to mix it up a bit, he had more meat. Lamb chops. Rib steak fried in plenty of butter and served with rich, creamy sauces. On the one hand, it was great for him, especially since the constant, nagging feeling of hunger he had experienced while dieting was now replaced by the pleasant sensation of a full stomach. On the other hand, Skaldeman knew that his body would not last long under this sinful dietary regimen.

But then came the great surprise: after a few weeks of eating like this, Skaldeman got on the scales and for the first time he did not hear that familiar, fatal overload alarm. Miraculously, the display came to a stop before exceeding its maximum capacity. A year later, Skaldeman was slim and feeling as fit as a young antelope.[3]

Insulin resistance: *the* metabolic disorder of the well-fed world

Most nutrition researchers and medical professionals would dismiss Skaldeman's metamorphosis as a 'nice anecdote'. Interesting perhaps, even heartwarming, but not proof of anything — indeed, generally misleading, as that kind of meat-heavy, Atkins-like diet is not exactly known as the healthiest way to eat. So why am I telling Skaldeman's story here? Have I already forgotten my own exposition on Atkins and animal protein?

The answer is that I'm not telling Skaldeman's story because I think his change of diet is generally a good or recommendable one. On the contrary, current scientific knowledge would indicate that it's not. For me, Skaldeman's diet is not one to be emulated, and certainly not for everyone across the board. No, I would say Skaldeman is an extreme case. But sometimes, extreme cases teach us something important about the situation in general by illustrating the core of a basic phenomenon in an extravagant and therefore eye-catching way. That's why I'm telling Skaldeman's story — because his experience teaches us something pivotal. But what, exactly, does it tell us?

I'd probably not have taken Skaldeman's metamorphosis quite so seriously if there hadn't been such an increase in recent years in the number of discoveries and insights that substantiate his experience in principle. These recent findings show that Skaldeman's story is far more than just an anecdote. It begins with the following simple, almost obvious observation, which is nonetheless very important in practice: when different diets (Atkins, Zone, low-fat, etc.) are tested against each other, it becomes clear that the success of any particular diet is an extremely individual matter. Irrespective of which diet is tested, there will be some test subjects who succeed in losing weight on it. They might lose 10, 20, or as much as 30 kilos in the space of a year, and in exceptional cases like that of Skaldeman, even more. Their bodies react extremely well to the new diet — as if it had been designed especially for them. But then the *same* diet has absolutely no effect on other people. Even worse, some actually gain weight on the diet!

Scientists at Stanford University in California examined this aspect over many years in several studies.[4] In their original experiment, researchers randomly placed more than 300 obese women on one of four different diets, ranging from Atkins to low-fat. Their conclusion after one year was that although the Atkins diet was found to be the most effective *on average*, some women who followed it lost no weight at all. A few women were in fact fatter after following Atkins for a year than they were at the start of the study. There was a similar picture in the results of the low-fat diet. Indeed, this phenomenon turned up with all the diets they studied.

Surprised? Okay, but now it really gets interesting. Since the women's blood levels were all different when they were measured at the start of the study, the researchers decided to examine them again and compare them with the women's success at losing weight. And what they found was remarkable. Women with a higher sensitivity to insulin responded better to a low-fat diet than to a low-carb one — an effect that has also been encountered by other research teams.[5] The results were different, and indeed opposite, for women with a pronounced resistance to insulin. It was as if these women's bodies were totally unable to cope with a low-fat diet. These women hardly lost any weight if they happened to be assigned to the low-fat group. Their bodies reacted far better to a low-carb diet. This link has also been observed repeatedly in a number of investigations at various universities (overall, it's even clearer and more consistent than the low-fat effect among women with greater insulin sensitivity).[6] In summary, we can say that when a body is sensitive to insulin, it is able to process carbohydrates without any problems, even in great quantities. But when it is resistant to insulin, carbohydrates become problematic. Thus, in simplified terms, insulin resistance is a kind of carbohydrate intolerance.

A closer inspection of these research findings reveals that the insulin-resistant test subjects who were put on a low-fat, carbohydrate-rich diet didn't even manage to *adhere* to it. They just abandoned it! It seems such a diet simply went totally against the grain with them.[7] What might be the reason for their rejection? What do you think? What's the usual reason people fail to stick to a diet? Of course, there can be any number of reasons, but usually the deciding factor is hunger, pure and simple. Perhaps a low-fat diet is particularly unsatisfying for certain people because it leaves them feeling especially hungry for some reason?

Before we look more closely at this question, let's just recap what the above means: there are some insulin-resistant people for whom a low-fat diet is doomed to fail from the outset. Low-fat necessarily means that most of the nourishment in a diet must come from carbohydrates. As we know, for many years the 'official' dietary recommendation has been to avoid fat and instead tuck in to carbohydrates, in the form of bread, pasta, rice, or potatoes, among other things. So those who, like Sten Sture Skaldeman, dutifully follow the generally accepted dietary guidelines but are resistant to insulin are simply banging their heads against a brick wall with their well-intentioned attempts to shed some kilos. They are engaging in precisely the behaviour that will not work with their body, for reasons I'm about to examine. It's not until such people dare, as Skaldeman did, to do what is 'forbidden' and switch to that notoriously 'dangerous' low-carb diet that their bodies begin to react and the rolls of fat begin to melt away.

The bitter irony is that it is *precisely* overweight people who tend to be affected by insulin resistance.[8] The reason for this is that after constant overeating for a long period of time, our fatty tissue becomes unable to store any more fat (although there are considerable differences between individuals in terms of how much fat their cells can cope with). If all the storage space in your home is full to overflowing but you continue madly shopping online and accumulating all those irresistible products from the internet, the rest of your home will eventually have to be sacrificed to make room for the flood of fresh purchases. Sooner or later, the entire house, including the living spaces, will be jam-packed full of stuff. A similar process takes place in our bodies, when the usual capacity for storing fat (in the fatty tissue just under the skin) reaches its limit.

Surplus calories are now stored in the form of fat in places where fat doesn't belong. The fat is increasingly deposited in the abdominal region and inside the cells of our organs — for example, in liver or muscle cells.

Intra-abdominal fat secretes inflammatory substances that can alone cause the body to become resistant to insulin. This is compounded by the fat stored in the cells of the liver and the muscles. That fat disturbs the signalling pathways inside the cells, including the signalling pathway for insulin, a disturbance that also leads to insulin resistance. (Thus, to a certain extent, initial weight gain due to overeating is the price we pay for not immediately becoming insulin resistant and developing diabetes. The extra calories are kindly 'tucked away' by our healthy fatty tissue. But the more those classic fat-deposit sites are stretched to capacity and beyond, and can no longer absorb those calories, the closer the body comes to insulin resistance and eventually diabetes.)

When the entire body becomes 'fattened', the result is insulin resistance. As obesity increasingly becomes the norm around the world, insulin resistance is also increasingly becoming an inherent part of that normality: insulin resistance is quite simply *the* metabolic disorder of a well-fed world. That means insulin resistance is no longer an exotic exception, but is becoming the rule. This, along with the fact that its consequences are so serious, is why I mention insulin resistance so often in this book.

To sum up: the fatter we are, and, in particular, the bigger our waist circumference is, the more insulin resistant we usually are. In other words — and this is the crux of the matter — precisely when we are most in need of a diet, our bodies fail to respond favourably to the traditional dietary recommendations, responding better instead to the low-carb regime that's still so frowned upon by most nutrition experts! Indeed, we are warned about the dangers of the dietary program that works best in the situation we find ourselves in.

In fig. 5.1, I once again give a summary of this interdependency.[9] Insulin-sensitive people who want to lose weight cope better with a low-fat diet. Insulin-resistant people, on the other hand, do better on a low-carb diet. So those who want to lose weight need to answer the crucial question of what their insulin status is. Are the cells of their

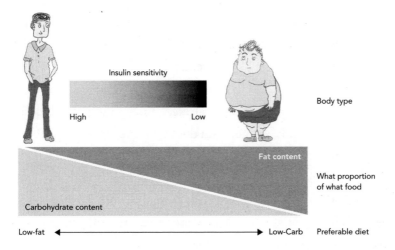

body insulin sensitive or insulin resistant?

Fig. 5.1 The external indications for our cells' level of insulin sensitivity that we can best rely on are body weight and waist circumference. Insulin resistance is quite rare in people of normal weight with a waist circumference of less than 100 centimetres. That means those people's bodies typically react well to the hormone insulin. In such cases, a low-fat, high-carb diet is best. The more surplus kilos we're carrying and the more our waist circumference expands, the greater the risk of insulin resistance. In such circumstances, the best diet is low in carbohydrates and rich in fat. (You may be asking whether the guy on the left is in need of a diet at all. And you are right: he probably doesn't. Please don't take the heavily exaggerated cartoon figures in the illustration too literally, but consider them to be the two extremes of a continuum.)

The question can only be definitively answered by a doctor or in a suitable laboratory. However, there are some 'external' indications that suggest insulin resistance might be at play. Your risk is increased if:

- As already mentioned, you are overweight and do not take enough exercise.[10]
- The circumference of your waist is more than 100 centimetres (measured at belly button level).[11]
- You have close relatives (parents, siblings) with type-2 diabetes. Type-2 diabetes is usually a direct consequence

of insulin resistance — at this stage, the body is no longer able to compensate adequately for the insulin resistance by providing enough (additional) insulin to regulate the amount of sugar in the blood. As a consequence, blood-sugar levels are permanently too high.[12]

- You suffer from high blood pressure, e.g. if you repeatedly have values of 140/90 mmHg and higher. Reasonably priced blood-pressure gauges are widely available, and checking your blood pressure regularly is a good idea.
- Your doctor finds the triglyceride levels in your blood are raised when measuring your blood lipids (from 150 milligrams per decilitre or 1.7 millimoles per litre) and your HDL levels are too low (roughly speaking, lower than 40 milligrams per decilitre, or 1 millimole per litre).[13]

You can measure your own blood-sugar levels. Diabetics do it all the time, and blood-sugar meters are also very reasonably priced these days. If your blood-sugar level is unusually high — on an empty stomach or after a high-carb meal — that may also be a sign of insulin resistance. Your cells can only respond very weakly to the insulin commands, resulting in more sugar circulating in your blood.

You may be asking how your blood-sugar levels can be raised at all when you haven't eaten anything — for example, in the morning before you have breakfast. After all, your stomach will have been empty for hours. The answer lies in the activity of your liver. When your blood-sugar levels fall during the night because you haven't eaten anything, your liver springs into action and pumps glucose into your bloodstream to make sure your brain in particular is constantly supplied with energy. Insulin inhibits the production of glucose by the liver, to prevent the liver from overdoing it. In this way, insulin also regulates our blood-sugar levels through the night. But if your liver becomes fatty and insensitive to insulin, the hormone loses that inhibitory effect, and the result is that the level of sugar in the blood is (too) high, even on an empty stomach. Thus, a raised level of sugar in the blood in the morning is a sign of an insulin-resistant liver.

Our blood-sugar level before breakfast in the morning is usually between 70 and 100 millilitres per decilitre (or, in millimoles per litre, between 3.9 and 5.6). Any value above 100 isn't ideal. A value of 126 (7 mmol/L) or higher means an official diagnosis of diabetes. Note that there are stricter criteria for pregnant women — they are said to be officially suffering from gestational diabetes with a value of approximately 92 to 95 milligrams per decilitre.

Everybody's blood-sugar level increases following a (high-carb) meal, but if it rises to 200 (11.1 mmol/L), the situation is definitely out of hand. This is a signal that the muscles have become insulin resistant and are no longer absorbing sugar from the bloodstream. If you find that your blood-sugar level is raised, you should consult a doctor to find out what's causing it and what you can do about it.

Fat as an alternative fuel

While we are still young and slim and dynamic, the cells of our body are usually still highly sensitive to the hormone insulin. That means our body has no trouble utilising the carbohydrates we eat and which then enter our bloodstream as a source of energy; that is, to 'burn' them, so to speak. Our carbohydrate metabolism is intact.

As we get older, however, that sensitivity to insulin typically declines, and the result is that we are unable to tolerate as much carbohydrate in our later years. This means age is a further risk factor for insulin resistance.

In addition, you need to know the following: your fasting blood-sugar level means your blood contains little more than a teaspoonful of glucose (about five grams). When you think that those few grams of sugar are dissolved in the five to six litres of blood in your body, it becomes clear blood is not a particularly sweet nectar. (To be clear: when we speak of 'blood sugar', we always mean glucose.)

As soon as you eat a couple of slices of toast or a plate of potatoes or pasta, the situation temporarily changes, and many times that amount of sugar flows into your bloodstream. These carbohydrates consist mainly of starch, which means they're made up of long chains of glucose molecules attached together. The long sugar chains are broken down into individual glucose molecules in the intestines so that they can be absorbed by the gut. Your blood is now highly enriched with simple sugars. However, your body is averse to having either too much or too little sugar circulating around it. One of the reasons excessive glucose is harmful is that the sugar in your blood has the propensity to 'stick to' all sorts of structures, especially the protein structures in the body.

For example, glucose molecules can stick to haemoglobin, which is the substance that makes your blood cells look red. Your doctor can also measure the proportion of 'glycated' haemoglobin in your blood, which is expressed as a value known as HbA1c. It's important to know your HbA1c value because it provides information not about your current blood-sugar levels, but about your blood-sugar situation over the previous two to three months. An elevated value (approximately 6 per cent or higher) indicates that regulation is impaired: the blood-sugar level is permanently too high, the body is 'sticky' on the inside, and that is also a type of ageing.

Incidentally, some spices and other plant substances (such as the previously mentioned flavonoids found in dark chocolate) can boost the insulin sensitivity of our bodies and have a positive influence on blood-sugar levels. One striking example of this is cinnamon. Regularly eating cinnamon lowers your blood-sugar level, also resulting in a lower HbA1c value. (Note: be sure to choose 'true' cinnamon from the Sri Lankan Ceylon cinnamon tree rather than Chinese cinnamon from the cassia plant, as the latter contains large amounts of coumarin, which can be poisonous if consumed in great quantities.)[14]

Because glucose molecules behave so aggressively when they are in the bloodstream, the body acts immediately when their numbers begin to rise by trying to remove the surplus glucose molecules from the blood and stashing them in the cells of the body. Conversely, our brain starts to panic when the amount of sugar in our blood falls below a critical level because it requires a continuous supply of energy to keep functioning and to survive at all, and that continuous supply cannot be secured by fat as an energy source. Our brains depend on a steady supply of glucose. This is why our bodies closely monitor our blood-sugar levels.

When insulin is secreted by the pancreas and goes knocking on the doors of the cells of our body, the cells increasingly open up to let in sugar molecules that are in the blood. That happens every time we eat a meal rich in carbohydrates (consisting of glucose). However, insulin is also a fat-storage hormone. The biological logic behind this is as follows: when there's a lot of glucose circulating in our blood vessels, it normally means that we have just eaten. When our blood-glucose level, and therefore our blood-insulin level, is high — that is, when the energy-supply situation looks good — we don't need to burn fat as an additional energy source. Therefore, insulin blocks the burning of fat. Since insulin resistance in the muscles and the liver results in raised blood-insulin levels, our fatty tissue stubbornly holds onto its fat if we are insulin resistant (until at some point, it stops reacting to insulin itself). Even with well-stocked fat reserves, we're unable to tap into this energy resource — like money in a fixed-term deposit account that you can't just take out and spend. This may be the explanation for how we can still feel hungry when we have plenty of rolls of fat in store. The fat in our body's cells makes them resistant to insulin, which causes increased insulin levels, which, in turn, make it impossible for us to utilise that fat as a source of energy since insulin blocks the burning of fat for energy.

The precise connections are, as always, very complicated and far from fully decrypted by scientists. However, a basic assumption is that every time an insulin-resistant person tucks in to a load of carbs — bread, fried potatoes, a mountain of rice — it's like pouring oil on the fire, as the flood of carbohydrates causes the level of sugar in the blood to rise, in turn causing the already high levels of insulin in the blood to rise even further.

To a certain degree, the protein we eat also leads to an increase in insulin levels. The only essential nutrient that is metabolically less dependent on insulin is fat.[15] This might be the reason why the bodies of insulin-resistant people respond better to a low-carb, high-fat diet: their bodies process fat as a fuel more efficiently than carbohydrates. For them, fat is a kind of 'alternative energy source', which can be utilised efficiently despite insulin resistance.[16]

Studies show that for people with insulin resistance, avoiding carbohydrates and switching to fat as the main source of fuel alters their entire metabolism for the better.[17] Their chronically fluctuating insulin levels finally come to rest, the level sinks, and — according to the latest speculations — insulin-led fat storage ends. The fat can finally escape the fatty tissue and become available to the rest of the body. It's like finally gaining access to the money on that fixed-term account. Despite the large amount of fat consumed — indeed precisely because so much fat is being consumed — the kilos fall away because of the lower insulin levels. Another beneficial result is that the surplus energy from the fat is set free, and that constant feeling of hunger disappears.

Incidentally, this recovery process can be dynamically supported and accelerated by exercise. The fact is that when we start to exercise, our cells 'automatically' start absorbing more glucose from the bloodstream, and they do so *completely independently of insulin*. Depending on the intensity of the exercise, our muscle cells can absorb up to 50 times as much glucose as normal![18] Another effect of exercise is to make the body more sensitive to the various signals that tell us when we are full following a meal (of which insulin is one of the most important). So there's no doubt that exercise helps with losing weight.

But exercise alone is not enough![19] That's because exercising makes you hungry and therefore likely to eat more.[20] In one study, British behavioural neuroscientists had a group of test subjects take part in an exercise program over a number of weeks. This revealed that the people who lost a substantial amount of weight during the program ate more fruit and vegetables during that time. By contrast, others — for whatever reason — ate *less* fruit and *fewer* vegetables, and tended to eat more junk food, with the consequence that they barely lost weight despite the exercise program.[21] To put it another way, diet is key to losing weight, even when you exercise intensively.

I myself used to believe that the reason I could stuff myself full of anything without gaining weight was because I ran so much. It's a misconception that often backfires. Yet if exercise is combined with a healthy diet, it can indeed lead to very profound changes: a feeling of inner balance, increased vigour, lower blood pressure, less stress, better sleep (which in turn further lowers stress, and so on). And yes, in addition to all that, or in connection with it, exercise turns out to be a highly effective treatment for insulin resistance — although the phrase 'highly effective' is an understatement. Coupled with weight loss, exercise has been shown in reliable studies to be a *better* way of combating insulin resistance and protecting against diabetes than metformin, which is currently the most widely prescribed medication for treating diabetes.[22]

Low-carb beyond Atkins

The low-carb community is right in many respects: the bodies of people who are insulin resistant have a certain intolerance to carbohydrates. It is better not only to cut out sugar, but also to cut down on all carbohydrates — bread, pasta, rice, and potatoes. Since insulin resistance is on the rise due to increased rates of obesity, it's not unlikely that the low-carb movement will further increase in popularity.

I believe the low-carb fans' biggest weakness is that many of them are so fixated on the hormone insulin that they fail to see the whole picture. Their main concern is to lower their insulin levels. But they ignore the fact that there are also other factors that play a part in a healthy diet.

I have met many (smart) proponents of the low-carb diet who, as you might expect, do not hold back on meat, bacon, and eggs, just like Sten

Sture Skaldeman. Since such foodstuffs contain little to no carbohydrate, they are 'good'. Full stop. And when you cautiously mention studies that indicate eating too much animal protein, and in particular processed meat products, might not be totally unproblematic (quite apart from the fact that protein also stimulates insulin production), they just argue it away, some more ably than others. For many low-carb fans, carbohydrates are so morbidly evil, so abysmally awful, that anything else appears harmless, or even like a healing medicine, by comparison. And that attitude may be justified to some extent in extreme cases. Is Skaldeman healthier than he was 20 years ago? Obviously, yes. Was it a good idea for him to completely change his diet? Clearly it was. Is his diet ideal? Certainly not. Do insulin-resistant people have no choice but to follow his example? No.

It's time now for a summary. If you want to lose weight and suspect or have been told by your doctor that your sensitivity to insulin is reduced, you should definitely try a reduced-carb diet. The best strategy is to try it for two or three weeks to see how your body reacts. In my case, I can get back down to my ideal weight relatively easily after a bout of binging (usually due to stress or travel), by simply reducing my intake of carbohydrates strictly for a time. I don't follow the Atkins diet. And anyone wanting to try out a low-carb diet does not necessarily need to follow the Atkins diet — indeed, would probably do better not to. There is another, healthier way:

- Cut out processed meat in the form of sausages, bacon, ham, salami, and hot dogs. The occasional piece of wild game meat, grass-fed steak, or free-range chicken is okay; however, shellfish, and oily fish in particular, are far better.
- Butter, cream, and coconut oil are allowed; however, generous use of high-quality olive oil (several tablespoons a day) is better. Two tablespoons of olive oil a day not only help overweight type-2 diabetics to lose weight, but also reduce blood sugar and AbA1c levels within just a few weeks.[23]

- You can also try an 'MCT oil' to further boost fat burning (MCT stands for 'medium-chain triglycerides', which is a fatty acid of medium length, on which more in the later chapters on fat). Coconut oil is often described as an MCT oil, but its MCT content is only 15 per cent. There are, however, 'pure' MCT oils (mostly extracted from coconut oil) that are made up entirely of those beneficial medium-length fatty acids. Initial research results indicate they can help with weight loss and increase insulin sensitivity (my cautiousness here stems from the fact that the findings are still very recent and, especially, from the fact that we're dealing here with a processed product).[24] The recommended daily dose is 10 to 20 grams, which is equivalent to about two to three tablespoons a day. One major factor in the rise in popularity of MCT oils is 'bulletproof coffee', with which you can treat your fatphobia as soon as you get up in the morning. The recipe: a cup of coffee, a tablespoonful of butter from meadow-grazed cows, and one to two tablespoons of MCT oil. Mix well with a hand blender. If that doesn't wake you up in the morning, nothing will!

- All vegetables are extremely welcome. Eat as many as you can, with the strict exception of potatoes, which raise the blood sugar level too quickly (on which more in the next chapter).

- Eat large salads, garnished with plenty of seeds. Personally, my evening meal consists of 'just' salad — lamb's lettuce, cos lettuce (romaine lettuce), rocket (arugula), and so on — several times a week, sometimes with prawns, falafel, porcini or chanterelle mushrooms. Just a little tip on the side: turmeric (a kind of yellow-orange, ginger-like root) makes a great addition to both falafel and salad dressings. It's also known to improve our cells' sensitivity to insulin.[25]

- Eat more healthy sources of protein and fat, in the form of beans, lentils, chickpeas, nuts, olives, and avocados (although beans, etc. also contain plenty of carbohydrates, they are still healthy, as we will see in the next chapter).

- All edible mushrooms (button mushrooms, shiitake, etc.) are highly recommended.
- Cheese is okay, as are eggs in moderation (I only buy organic, free-range eggs. As rule of thumb, seven eggs a week, i.e. a maximum of one per day *on average*). One of my personal favourite low-carb lunches is a 'Caprese salad' (tomatoes, mozzarella, fresh basil leaves, a little balsamic vinegar, and plenty of olive oil — best eaten with a spoon!).
- Greek-style yoghurt is an excellent choice for dessert — for example, with chia seeds or linseeds (both of which are high in fat, incidentally) as well as possibly some fruit. Berries are particularly good. Blueberries, for example, are known to increase insulin sensitivity.[26] Apples contain a touch more sugar, but are still a good choice. Apples — especially their peel[27] — are a concentrated source of phlorizin, a plant substance that inhibits the absorption of glucose by the small intestine. This results in less glucose entering the bloodstream, which correspondingly inhibits insulin response.[28]
- Other beneficial foodstuffs: green tea[29] and, as mentioned before, *very* dark chocolate (the darker the chocolate, the less sugar it contains).[30] Both have been proven to boost insulin sensitivity.
- Last but by no means least: a small glass of *dry* wine with your evening meal is allowed; more than allowed, in fact.

The final comment isn't a joke. In a unique experiment, 200 diabetics were divided into three groups. One group was given red wine, another was given white wine (both dry varieties, which means they were low in sugar), and the unlucky third group was given mineral water. For the next two years(!), they were asked to take 150 mL of their assigned drink with their dinner every evening. At the end of the two years, only the two wine-drinking groups showed an improvement in their blood-sugar regulation, with the most pronounced effect in the white-wine group (their average fasting blood-sugar levels sank by no less than 17 milligrams per decilitre compared to the mineral-water group).[31] For more on the subject of alcohol, see chapter 7 on drinks in general.

One final remark: insulin resistance and the carbohydrate intolerance it brings with it are not all-or-nothing phenomena. Not all of us are Sten Sture Skaldeman, who really would be well-advised to reduce his carbohydrate intake to a minimum for the rest of his life. For most of us, our insulin resistance is certainly not as pronounced as his, which means that we can tolerate more carbohydrates than he can.

In addition, when you lose weight, it's not only the classic fat deposits that melt away. Intra-abdominal fat also disappears, along with the inflammatory substances it secretes, as well as the surplus fat in the muscles, liver, and other cells of the body. That means your cells regain their sensitivity to insulin. The more fat deposits you lose, the more sensitive your body becomes to insulin, and the more its intolerance to carbohydrates abates. This occurs to such an extent that a radical (i.e. low-calorie) diet and the resulting major loss of weight can often reverse and cure even a substantial case of diabetes.[32]

Most of us can then begin successively introducing more carbohydrates back into our diet after a few weeks or months. The best person to judge the optimum amount of carbohydrates in your diet is you. And the only way to find that out is by trial and error. Ask yourself how you feel after starting to eat bread and pasta again. Does it make you hungrier than before? And monitor your weight after each change in your diet to check if the kilos are creeping back. Does your weight remain stable? This requires a certain willingness to experiment and an ability to monitor your own progress. For me, by the way, experimenting like this is not just an interesting exercise, but also a comforting one. It's reassuring when your experiences let you know that you aren't totally at the mercy of your body weight, but rather, with some practice, you can take back control of your body. However, it is important, especially when reintroducing carbohydrates into your diet, to concentrate on healthy sources of carbs. We will explore which those are in the next chapter.

CHAPTER 6

Carbohydrates III: how to recognise healthy carbohydrates

The four crucial quality criteria

For most of us, the crucial thing from a health point of view is not the relative amount of carbohydrates we eat, but the *type* of carbohydrates. What is it that makes some carbohydrates healthier than others? Here are four crucial criteria:

1. **Solid or liquid.** Since I've already dealt with this topic in chapter 4, all that's needed here is a brief recap. Whole fruit is always better than juice. Apples and apple juice really are two very different creatures. Firstly, the whole fruit contains more nutrients. Secondly, the sugar in whole fruit enters the bloodstream less quickly, as it comes in an intact structure with dietary fibre (also think of such substances as phlorizin, which is found mainly in the peel of an apple). Thirdly, eating whole apples or oranges, for example, is less likely to tempt you

to overdo it. I can happily wash down the equivalent of eight 'apples' or 'oranges' in the form of squeezed juice.

2. **Level of processing.** The more natural and original the state of the food we consume, the better it is for us (exception: in the case of some vegetables, the beneficial plant substances they contain can sometimes only be released and made available for our bodies by chopping and heating. This is the case, for example, for lycopene, the red pigment contained in tomatoes).[1] In this respect, the processing of grain is extremely relevant. As you will see, there's an enormous difference between a wholemeal loaf, a brown loaf made of finely ground whole grains, and white bread.

3. **Dietary fibre.** Another rule of thumb for considering high-carb foodstuffs is the question of their fibre content compared to the overall carbohydrate content (dietary fibre, which makes up part of a plant's cells, is made up of carbohydrates that our bodies can't digest). You should try to get as much fibre in your diet as possible. It's good to aim for a digestible-carbohydrate-to-fibre ratio of less than 10 to 1, which is when 10 grams of carbohydrates provides at least 1 gram of fibre.[2] Even better is a ratio of less than 5 to 1. Often — but, unfortunately, not always — the amount of fibre contained in food is included in the nutritional information on the packaging. For example, 100 grams of the white basmati rice I usually use contains 78 grams of carbohydrates, but only 1.4 grams of dietary fibre. 78 divided by 1.4 equals 56 — so for every gram of fibre there are 56 grams of carbs. The beluga lentils I like are much better: they provide 17 grams of fibre for every 41 grams of carbohydrates. 41/17 = 2.4 — that's a ratio of well below 5 to 1, which is just ideal. This is why I much prefer lentils to rice (another reason is that rice is often contaminated with arsenic).[3]

4. **Glycaemic index (GI).** This is a measure of how quickly carbohydrates can be digested by the body. Rapidly digested

carbohydrates cause harmful spikes in blood-sugar and blood-insulin levels. The glycaemic index is a criterion you should only use when in doubt about which of two foods from the same category is better. For example, rice: the white basmati rice I mentioned earlier is low in fibre, but at least it releases its carbohydrates into the bloodstream slowly in comparison to jasmine rice, for example, whose carbs enter our blood almost at the speed of light. Even if basmati rice isn't ideal, it's better than jasmine rice.

Now, let's take a closer look at these four criteria. The aim is to be in a position by the end of this chapter to assess for yourself how healthy any high-carb food is.

Does bread make you fat and sick?

Let's start with poor old bread. What happened to the good reputation bread used to enjoy? It's a painful development, especially in a bread-loving country like Germany, with its more-than 1,000 different types of bread! (The village I live in is small, but, astonishingly, has four bakeries.) If you read popular dietary-advice books such as *Wheat Belly* or *Grain Brain*, you might be forgiven for developing a fear of ever entering a bakery again, let alone actually biting into a slice of bread. Wheat, bread, and grain of any kind, they warn, make us fat, lethargic, stupid, and very, very sick. The main culprit, they claim, is the demonic protein compound called 'gluten'.

The thing about such advice books is that the critical view they present is *not* made up out of thin air. They contain elements of the truth, but go off track as a whole. The fact is that most of us can eat a couple of slices of bread a day with a clear conscience. However, it should be the right kind of bread. I advise against white bread (including rolls, baguettes, pretzels, and croissants). I know it tastes great, but I now treat white bread as I would a sweet treat. And that's what it basically is. White bread is, as the Belgian doctor Kris Verburgh puts it in his excellent book *The Food Hourglass*, 'what remains when you squeeze out every mineral, fibre, and nutrient from bread'.[4]

Every loaf of bread begins with the grain, whether that is wheat, rye, or spelt. The grain is ground, which is similar to squeezing fruit for juice — and thus the grain can be 'ground empty', with most of the beneficial substances falling by the wayside. What remains is almost exclusively made up of starch, which is pure carbohydrate in the form of long, bleak chains of glucose (see fig. 6.1).

Have you ever noticed that the packs of flour you buy in the supermarket come in different types? Some is as white as cocaine and about as nutritious, containing perhaps 405 milligrams of mineral nutrients per 100 grams of flour; the flour has been ground empty and what remains is nothing more than just the pulverised endosperm. Another type has 1050 milligrams of mineral nutrients per 100 grams,

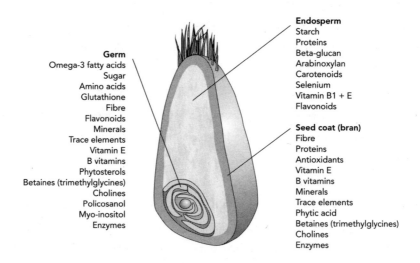

Germ
Omega-3 fatty acids
Sugar
Amino acids
Glutathione
Fibre
Flavonoids
Minerals
Trace elements
Vitamin E
B vitamins
Phytosterols
Betaines (trimethylglycines)
Cholines
Policosanol
Myo-inositol
Enzymes

Endosperm
Starch
Proteins
Beta-glucan
Arabinoxylan
Carotenoids
Selenium
Vitamin B1 + E
Flavonoids

Seed coat (bran)
Fibre
Proteins
Antioxidants
Vitamin E
B vitamins
Minerals
Trace elements
Phytic acid
Betaines (trimethylglycines)
Cholines
Enzymes

Fig. 6.1 A grain of cereal consists of a starchy endosperm, which contains a few vitamins, minerals, and other substances. Most of the beneficial substances, however, are to be found in the seed coat (bran) and the germ (or embryo). When the whole grain is ground into white flour, the multilayered seed coat and the germ are removed. All that remains is the nutrient-poor and low-fibre, but energy rich, endosperm. Here are a few figures: grinding grain into white flour leads to the loss of: 58 per cent of its fibre, 83 per cent of its magnesium, 79 per cent of its zinc, 92 per cent of its selenium, 61 per cent of its folic acid, and 79 per cent of its vitamin E.[5]

and has a light brown tinge to it; this flour still contains parts of the outer seed coat (bran). Wholegrain flour, as the name implies, uses the whole of the grain and has the maximum possible vitamin, fibre, and mineral content. Wholemeal bread is (for the most part) made from wholegrain flour — irrespective of which type of grain the flour comes from.

The lady behind the counter at my local bakery insists that spelt is 'much healthier' than wheat and other grain, which has always irked me, a big fan of rye bread. I don't know where she got that information from. There are virtually no scientific findings to back it up. Personally, I think it's just a myth. However, there are any number of studies that indicate that the important thing — irrespective of the type of grain — is that the flour used was wholegrain rather than one with most of its nutrients ground out. (Unfortunately, these days, many of the people working in bakeries don't know what 'wholegrain' really means, and they also often have no idea what's in the bread they sell because it's no longer baked on the premises. Don't let such 'bakers' confuse you! PS: I now bake all my own bread, not least because of all the unnecessary additives typically found in industrially baked bread. It's surprisingly easy and enjoyable. You can find the recipe I follow in this footnote.[6])

It's not just the amount of nutrients contained in various foods that's important. As you've already seen, the digestion process is also a factor. The more the grains are milled, the finer the resulting flour particles will be, and fine particles are particularly easy and quick to digest. Pulverised in this way, the carbohydrates find their way into the bloodstream at lightning speed. The same is true of bread made from finely ground wholemeal. So the ideal choice is a coarse, granular wholemeal loaf. The intact seed coat acts as a beneficial physical barrier. It encases the carbohydrates in the endosperm, rendering them less easily accessible to our digestive enzymes to break them down into their individual components — glucose molecules.

White bread, by contrast, brings together all the negative aspects. There are no seed coats, very few nutrients, and the flour particles are very fine. White bread could be seen as the embodiment of one of the central problems of our affluent diet, which is that it leaves us simultaneously undernourished and over-nourished. It isn't lacking in energy; we certainly get enough of that from it. But it is often lacking in certain nutrients that our bodies need to keep running smoothly and to protect from premature degeneration. These nutrients include vitamins, such as the B-vitamin folic acid; minerals, such as magnesium and selenium; and beneficial fats, such as omega-3 fatty acids — all of them substances that are removed from grain by excessive milling. If most of the food we eat is high in energy but low in nutrients (white bread, white rice, white pasta, white sugar), we may fail to adequately meet our body's need for these essential substances.

This could be a contributing factor to obesity, as our body constantly sends out signals that it is lacking something (although we are eating so much!). We gorge and gorge ourselves, but since most of the beneficial substances our bodies need have been artificially removed from our food, we keep on eating till we have the feeling that our craving for those substances has finally been met. The result is a constant, nagging feeling of lacking something, a 'hunger for nutrients'.[7]

The nutrients contained in wholemeal products are not the only reason they are better at making us feel full. Fibre is also important. Oats and porridge are excellent in this respect. Porridge is made up of roughly chopped grains. Our digestive system has to work hard to break those guys down. To make oat flakes, the rough grain is flattened by rolling. This makes them somewhat easier to digest (albeit not to the same extent as milling) — it gives the oats a greater surface area, thus providing our digestive enzymes a larger area to 'attack'. In both cases, the oats contain large amounts of a kind of dietary fibre called 'beta-glucan'. When beta-glucan comes into contact with water, it turns into a slimy pulp. If you wait long enough, you can watch this happening in your morning bowl of muesli. But if you're too hungry to wait, it will happen at the latest inside your body. Once it reaches the gut, this glutinous gel both slows down the absorption of carbohydrates and inhibits the absorption of cholesterol. In this way, beta-glucan lowers the level of cholesterol in the blood.[8]

Our gut isn't able to digest fibre such as beta-glucan. This is why such substances used to be dismissed as pure 'roughage'. It's only in the past few years that scientists have realised how wrong that is. The fibre we don't digest is actually a feast for the previously mentioned guests we entertain in our gut, the gut bacteria (our 'microbiome'). And there's nothing better than having a gut full of satisfied bacteria! If you don't feed your gut bacteria enough fibre, they get their own back by beginning to eat away at the mucus membrane of your gut. And without that protective layer, your gut is left vulnerable to infection.[9]

Well-nourished gut bacteria also produce beneficial substances called 'short-chain fatty acids', such as butyric acid (or more precisely, butyrate). Butyric acid is a good source of energy for the cells of the gut, and even appears to act as a kind of medicine. For example, it inhibits inflammation and helps prevent cancer, among other things.[10]

A small portion of the short-chain fatty acids produced by our gut bacteria as they digest fibre passes through the gut wall into the bloodstream, eventually reaching the brain. There, they influence the neurons that control whether we feel full or not. In this way, our gut bacteria can tell our brains that they're full and we should stop feeding them. That means satisfying our microbiome's appetite for fibre is another way wholemeal products can help prevent obesity.[11]

In short, there's absolutely no reason to discourage anyone from eating wholemeal bread and wholemeal products in general. Quite the contrary, in fact. Scientists at Imperial College London and Harvard University recently analysed the data gathered in 45 studies on this topic. The result was that eating 90 grams of whole grain per day (for example, in the form of two slices of wholemeal bread and one bowl of cereal) lowers the risk of practically all conditions associated with ageing, from diabetes to cancer. The risk of cardiovascular disease was reduced by more than 20 per cent, and the mortality risk overall was also lower. In other words, eating modest amounts of wholemeal products makes for a longer, healthier life.[12]

The same is true, incidentally, of much-maligned wheat.[13] To be crystal clear: wheat per se is *not* bad. In his book *Wheat Belly*, the US cardiologist William Davis expounds on the way wheat products have left a trail of destruction around the world. This leads him to reject not only white bread, but all wholemeal products, describing them as a disaster for our health. In addition, as Dr Davis so subtly puts it, wholemeal products 'make us hungry and fat, hungrier and fatter than any other time in human history'.[14]

As we all know, the market is teeming with diet books that are so gaga it would be a waste of time to take them seriously. But books like *Wheat Belly* and *Grain Brain* are not among their number. These books are not the work of fools. The authors know their subject. They quote serious scientific studies. So how can they reach such false conclusions?

I think there are several reasons for this. I want to examine them a little more closely, because this is a typical example of why there's so much confusion and uncertainty surrounding nutrition research.

To start with, there *are* people who can't tolerate wheat and other types of grain, just as there are people who can't drink milk or eat peanuts. For example, wheat allergies are a thing, and so is gluten intolerance (the severe form of which is called 'coeliac disease'). Even if you don't have a medically diagnosed gluten intolerance, it may still be the case that your body has a hard time dealing with wheat or grain in general. Trial and error is the key! Cut grain out of your diet for a few days — or even better, a few weeks — and monitor how your body reacts. Just because something can't be diagnosed by a doctor, it doesn't mean it isn't real. When it comes down to it, your body knows many things your GP has no idea about. That said, however, the latest research indicates that most of us have no trouble dealing with gluten.

Gluten is the Latin word for 'glue', and this sticky substance is what makes pizza dough become nice and stretchy when it's kneaded — so much so that practised pizza makers can spin it into a disk on one finger. Gluten intolerance means the gut reacts rather unhappily when confronted with this protein. And what does 'reacts unhappily' mean? It means that, for some reason, the body's defence system misinterprets the protein structures as an enemy and launches an attack — in the form of an inflammation — whose main effect, unfortunately, is to destroy the cells of the gut. The symptoms can range from abdominal cramps, diarrhoea, and anaemia all the way to neurological conditions such as headaches and even motor disorders caused by appalling damage to the brain. Approximately 1 per cent of the population suffers from the coeliac-disease form of gluten intolerance, which can only be controlled by cutting out all gluten from the diet. That means no longer eating wheat, but also many other kinds of grain, including rye and barley.[15]

Please don't misunderstand me: a gluten-free diet is absolutely essential for people with coeliac disease. Furthermore, such a diet *can* be healthy for anyone, depending on what they do eat. Since most people are more likely than not to eat unhealthy white-flour products (white bread, pizza, pancakes, pastries, biscuits, and so on), many people are doing themselves a favour overall if they avoid (such) grain-based products. And if that junk food is replaced by vegetables, pulses, nuts, and fruit, that new, gluten-free diet really becomes healthy. Against this backdrop, it shouldn't come as much of a surprise when people who change their diet to a gluten-free one feel better, even when they don't actually suffer from a gluten intolerance.

However, if we're talking about *general* advice about nutrition, it makes more sense to distinguish between white-flour and wholemeal products. The fact is that most of us don't need to avoid wholemeal products. Indeed, doing so would actually be counterproductive if those products were to be replaced in the diet by foodstuffs that are far less healthy.

Anyone who has experienced a positive effect from giving up wheat or grain in general, and therefore harbours the suspicion that grain has a devastating effect on our health and wellbeing, will be tempted to seek to support their suspicion with scientific studies — but it's human nature to begin to perceive the results of such studies in a highly selective way. Eventually, confirmation bias makes it difficult to see anything that doesn't support one's own assumptions.

This tendency has a fatal effect on current nutrition research. Since there have been such a huge number of studies in this area over the past decades, it's now possible to find a study about almost any food you care to mention that decries it as a poison or hails it as a cure-all. After all, more than 250 nutrition studies are published *per day*. I could find pertinent 'US studies' to prove to you that eating broccoli regularly will kill you, and then, after giving you such a fright *(What? Broccoli now, as well?? Really???)*, I would dutifully sell you my sensational anti-broccoli diet. To put it in Darwinian terms, you could say that the sheer number of findings has created an intellectual environment that's extremely favourable for the spread of diet gurus.

How can we best deal with this situation? I think the appropriate strategy is to take as unbiased a view as possible of the overall research, rather than hastily and subjectively restricting yourself to one part of the picture. However, in view of the flood of studies, it seems that precisely that — taking an objective view of the overall situation — is almost impossible, or at least a feat of Herculean proportions.

Let me add a brief explanation. Myriad individual studies form the basis of our knowledge about nutrition, but they are not always reliable and are often contradictory. In order to wring a little more reliability out of this flood of studies, some research teams regularly pore over the soundest of the studies and collate their results. These summaries are known as 'reviews' or 'meta-analyses'.[16] That can give us a certain perspective to start with. In addition, the method used by such meta-analyses has been refined in recent years in studies that carry out a meta-analysis of the results of other meta-analyses — they could be considered meta-meta-studies. These super-studies give us an overall idea of the broader-perspective studies.

For example, a team of French researchers recently analysed the data and findings from all(!) meta-analyses and reviews on the topic of nutrition and health/disease published between 1950 and 2013. This was the first study of such size. The results are revealing — especially in the case of the 'controversial' issue of wholegrain products.[17]

Fig. 6.2 summarises the main results of that French meta-meta-study. The diagram shows how regularly eating foods from a certain category — for example, red meat, fish, or eggs, but also wholemeal products — is linked to the risk of developing the major diseases associated with old age, from cardiovascular disease and diabetes to cancer. Before we turn to the individual results: even a quick glance at the diagram shows that *none* of the foods investigated were found to be entirely protective or entirely harmful, if we can even speak of definite causal links at all. There are contradictory findings for every kind of food.

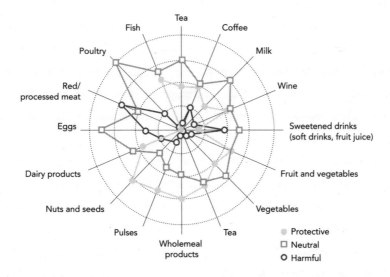

Fig. 6.2 The diagram shows the percentage of all large-scale analyses published between 1950 and 2013 that indicated a protective, neutral, or harmful link to the ten main diseases associated with ageing (cardiovascular disease, cancer, type-2 diabetes, liver disease, kidney disease, diseases of the digestive system and the skeleton, sarcopenia, brain disease, and obesity or adiposity). Taking wholemeal products as an example: almost 60 per cent of all large-scale analyses between 1950 and 2013 classify wholemeal products as protecting against at least one of the geriatric conditions considered. Nearly 40 per cent of all meta-studies came to a neutral conclusion (neither protective nor harmful). Only a very small percentage of studies (4 per cent) reached the conclusion that wholemeal products might be harmful. I'm not claiming that meta-analyses like this are the final word in questions about nutrition, but they do give us an impression of the general trend in the scientific assessment of certain food groups. This also helps to debunk senseless diet-hypes, like the current tendency to blame bread for everything, promoted by books such as *Wheat Belly* and *Grain Brain*.[18]

There are various reasons for this. To a certain degree, it might be due to the rather rough division of foods into undifferentiated groups (soft drinks are certainly more harmful than fruit juice; 'dairy products' include yoghurt, butter, cheese, and, in this case, sometimes milk itself; and so on). At a more basic level, however, these contradictions are a fundamental part of the way science works. Science is not a dictatorship. Data can be interpreted and evaluated in different ways. It's by gradually resolving such contradictions through critique and discussion that science constantly updates and corrects itself and thus makes progress.

Each individual study, however carefully it was carried out, has its own strengths and weaknesses. Observational studies can be wide-ranging and long-term — sometimes decades long — but it's then often difficult to pin down the crucial factor (think of the reason why coffee was considered so toxic for so long). Experiments carried out with two groups of test subjects following different diets are more scientifically rigorous, but they are by their very nature always limited in duration, which is an important factor when studying diets, whose effects are usually seen over a rather long period of time.

Then there's the fact that scientists are of course not infallible and sometimes make mistakes. Some researchers are heavily financed by the soft-drinks, milk, or some other industry, and, unfortunately, that dependency clearly colours some of their findings. For these and many other reasons, research can and will always yield contradictory conclusions.

Nonetheless, despite these contradictions, a general trend usually becomes apparent for each kind of food, concerning its effect on our health. It's this general trend that I've followed in this book. For example, the majority of the meta-studies (56 per cent) indicate that red and processed meats are harmful. There are many neutral studies, too, while 4 per cent still reach a positive conclusion about them. That means a person could praise red and processed meats to the high heavens as an elixir of life, if they focus only on those 4 per cent of studies that are positive about them and can find a reason to discount all the rest.

In the case of fish, the situation is reversed. 44 per cent of the studies were positive, 53 per cent were neutral, and 2 per cent negative. You really have to try hard to find anyone who categorises fish as unhealthy, but, of course, it *is* possible (and as we'll see, some types of fish are healthier to eat than others).[19] But you could even classify fruit as a danger to the public if you concentrated on the 2 per cent of studies that come to a negative conclusion about it, and disregard the other 98 per cent. Which brings us back to my anti-broccoli diet …

The case of wholemeal products is similar. Like fruit, such foods come off unusually well overall: no fewer than 60 per cent of the studies carried out since the beginning of the postwar period conclude that wholemeal products offer protection against many of the diseases associated with old age.[20] The rest are neutral. The meta-meta-study found only one research paper that categorised wholemeal products as negative. That's a positive outcome unmatched even by vegetables! The same is true, by the way, for the part of the study that looked at obesity: 40 per cent of all meta-studies reached the conclusion that eating wholemeal products is associated with *less* obesity, while 60 per cent were neutral. And how many studies do you think associated wholemeal products with weight gain? That's right: *not a single one*. So much for the 'wheat belly' hypothesis.

I don't want to give the impression here that the French meta-investigation is the be-all-and-end-all when it comes to nutrition research (it has its own weaknesses, which I'll return to when I look at milk). Yes, I believe it gives us a sound overall view of the general assessment of foodstuffs over the past decades — and that's all it does. However, in light of the disconcerting contradictions surrounding the issue of nutrition and also in light of the sometimes totally crackpot diet books on the market, an overarching evaluation of this kind helps in debunking and taking a more sober view.

The lead researcher in the team that carried out the meta-meta-study, Frenchman Anthony Fardet, was kind enough to make his raw data generally available, which is why I was able to reproduce the diagram in fig. 6.2. When I asked him whether the results of his work have changed the way he eats, he answered that they had indeed, and that he now eats more wholemeal products and follows a generally more plant-based and less animal-based diet.[21]

The glycaemic index

If you buy a blood-sugar meter of your own, you can monitor in your own body how strongly different kinds of food raise the level of sugar in your blood. You'll be surprised by what you see! Some foodstuffs that we consider extremely healthy 'basic foods' flood our bloodstream with so much of the monosaccharide glucose that we might as well just drink a glass of concentrated glucose solution. The most important example of such a food is the humble potato.

The speed of entry to bloodstream is what scientists refer to as the 'glycaemic index' or 'GI'. Its base is the measure of how eating a specific amount of pure glucose affects blood-sugar levels. Since blood sugar is actually glucose, this provides a reliable baseline reference value.

Let's assume you drink a glass of water with 50 grams of glucose dissolved in it. When you then test your blood-sugar level, you can watch it rise sharply over the next half an hour. It reaches a climax and then drops rapidly again, under the influence of insulin. You can see what this looks like as a curve on a graph in fig. 6.3.

The area below the curve can now be calculated, and the result gives us information about the rise in blood sugar in the time following ingestion of glucose (the timeframe is usually set at two hours following consumption). The greater the area below the curve, the greater the average increase in blood-sugar level. Since 50 grams of pure glucose gave us our reference value, that area is deemed to represent 100 per cent. By that definition, the GI of pure glucose is 100.

The blood-sugar effect of any food can be expressed as a curve like this, as long as it contains enough carbohydrates to cause a change in the amount of sugar in the bloodstream. Comparing the area below the curve with that of pure glucose results in the GI of that particular food.

The area below the curve for potatoes, for example, is 85 per cent the size of that for glucose. That means we can say the GI of potatoes is 85. (It's important to remember that this, like all GI values, is an average usually calculated from the results of around ten test subjects. Here, again, there can be large differences between individuals — more evidence for the fact that there's no such thing as the *one* diet that's perfect for everyone. This is why it's a bad idea to approach diet in a dogmatic way. It's far more sensible to experiment with various diets and observe your own body's reaction to them.)[22] In other words, 50 grams of carbohydrates eaten in the form of fried potatoes cause the level of sugar in the blood to rise almost as drastically as a drink of pure glucose solution. The glycaemic index of potatoes is unusually high for most of us, which doesn't exactly speak in favour of potatoes being a 'basic food'.

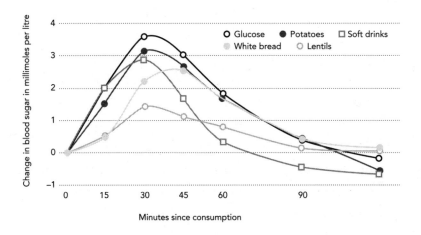

Fig. 6.3 The glycaemic-index graph shows how much the effect of various kinds of food on our blood sugar level can be — even though they contain exactly the same amount of digestible carbohydrate (50 grams of carbohydrate from each kind of food, *not* including any fibre they may contain)! For the sake of simplicity, the blood sugar level before consumption was set at zero, so the curve shows the change in blood sugar. Particularly striking is the difference between potatoes and lentils. The relatively low GI of soft drinks is due to the fact that about half the sugar they contain is fructose, which is mostly taken up by the liver and so never enters the bloodstream at all.[233]

Food	Glycaemic index (GI)
Glucose	100
Breakfast/bread	
Oats	55
Kellogg's cornflakes	86
Croissants	67
Rough-grain wholemeal rye bread	55
Sourdough bread	58
Fine-grain wholemeal wheat bread	74
White bread	71
Pretzels	80
Eggs	—
Home-made pancakes	66
Gluten-free pancakes (ready mix)	102
Fruit	
Apples	38
Bananas	52
Blueberries	53
Oranges	42
Pears	38
Strawberries	40
Vegetables	
Carrots	41
Tomatoes	—
Fried potatoes	85
Pasta/rice	
Spaghetti (white, cooked)	44
Wholemeal spaghetti (cooked)	42
Basmati rice (white, cooked)	58
Jasmine rice (white, cooked)	109
Dairy	
Low-fat milk	32

Full-fat milk	27
Low-fat natural yoghurt	35
Nuts	
Cashews	22
Peanuts	23
Walnuts	—
Drinks	
Orange juice	53
Coca-Cola	53
Beer	89

Fig. 6.4 The glycaemic index (GI) gives information about the speed at which the glucose molecules of a carbohydrate-rich foodstuff enter the bloodstream. The above is just a small selection (those who are interested in knowing the GI of other individual foods can find that information here: http://www.glycemicindex.com). The precise figures should not detract from the fact that there may be considerable differences between individuals (for example, white bread causes a significant spike in blood sugar *on average*, although, astonishingly enough, it has no such effect at all on *some people*!).[244] A GI of less than 55 can be considered low; between 56 and 69 is seen as medium; and anything above that is high. The reason I included the gluten-free pancakes (ready mix using buckwheat) here was to give an example of how 'gluten-free' doesn't necessarily mean 'healthier'. As a rule of thumb, I'd say the following: unless you have an intolerance to gluten, the main profiteers when you buy gluten-free products are the manufacturers. Many types of food (eggs, tomatoes, indeed most vegetables, almost all nuts, meat, all fats) contain very little to no carbohydrates, so they don't have an effect on blood-sugar levels — which means they have no GI. The GI of orange juice and cola is so low because a large proportion of the sugar they contain is in the form of fructose, which is absorbed by the liver and has no further impact on blood sugar.[255]

The blood-sugar curves for far more than 1,000 kinds of food have been measured over the years.[266] Comparing those curves shows that the GI isn't the only aspect in which potatoes stand out. Potatoes are also among the very few kinds of food that apparently provoke such a hefty insulin response in many people that it causes their blood-sugar level to sink a great deal two hours after consumption — an effect otherwise only associated with extremely sugary foods such as soft drinks and fruit juice (see fig. 6.3). This causes acute snack attacks, and especially cravings for rapid carbs, as our body tries to return its blood sugar to a normal level as quickly as possible.

All this might explain why potatoes are generally considered to be so fattening — a judgement that appears to be more than just a mere myth, and is supported by the results of large-scale observational studies by Harvard University and others (see fig. 0.1 in the introduction). One recent Harvard study even showed that eating potatoes in large quantities is associated with a slightly elevated risk of diabetes — presumably because of their high GI. Replacing three portions of potatoes per week with whole grains, by contrast, reduces the risk of diabetes.[277] Taking a view of the entire body of knowledge shows that potatoes aren't so terrible, although I myself am no longer a fan of the tubers and rarely eat them these days. Tip: if you really love potatoes, opt for the waxy ones as they don't cause such a rapid blood-sugar spike as the floury type.[28] Leaving potatoes to cool for several hours after cooking also has a beneficial effect. It creates a kind of 'resistant starch', which our bodies can no longer digest, but which is a good source of food for our gut bacteria.[29]

White rice is a similar case. Overall, rice's GI is also quite disappointing. Like eating a lot of potatoes, consuming large quantities of rice is associated with both obesity and an increased risk of diabetes.[30] However, some kinds of rice are worse than others. The GI of jasmine rice, for example, is so astronomically high (109) that it even exceeds that of pure glucose! Nutrition scientists speculate that this might be because a concentrated solution of glucose makes a short stop in the stomach after consumption, slightly slowing down its overall rate of digestion, while jasmine rice passes directly through the stomach and is digested at an unbeatable speed. As mentioned earlier, basmati rice is better in this respect. Basmati has a medium GI due to the slightly different composition of the starch it contains. I myself only eat rice in the form of sushi, and then only in small amounts. The advantage of sushi is that the seaweed ('nori'), the fish, and the rice vinegar lower its overall GI to below 50.[31] (Any kind of acidity slows down the rate at which the stomach and gut are emptied, which in turn leads to a lower

GI. This is equally true of vinegar and lemon juice, and it's the reason for the surprisingly low GI of sourdough bread.)

The reason I avoid eating rice in large quantities is actually the arsenic that accumulates in the grain. Rice is a plant that absorbs carcinogenic arsenic from water and the ground with the efficiency of a sponge. The plants are so talented at it that they can actually be used to 'detoxify' contaminated soil.[32] But the toxin then accumulates in the grain of the rice plant. In fact, rice often contains large amounts of arsenic, even wild rice and especially brown rice (white basmati contains a little less). This is no scaremongering myth — it has been proven on the basis of clear evidence. That's why I believe rice should not be considered one of the 'basic foods', but should be enjoyed occasionally as a side dish. Pregnant women and infants especially should avoid eating too much of it. Rice wafers, rice flakes, and other rice-based snacks are often especially contaminated: please stay clear of them! Also be cautious about giving rice pudding to small children. And rice milk is absolutely unsuitable for infants.[33]

The way it is cooked also makes a difference. For the longest time, I used the worst method possible: I boiled it in a saucepan, unwashed, with double the amount of water, until all the liquid boiled off. Preparing rice this way means all the arsenic remains in the grains (the same is true of rice prepared in a rice cooker). An Indian friend of mine showed me a much better method: First, rinse the rice thoroughly under plenty of flowing water until the water runs clear. Then boil the rice in a large pan with a couple of litres of water, as you would pasta. When the rice is cooked, sieve it and serve. The copious amount of water washes about half the arsenic out of the rice, and that arsenic ends up down the drain.[34]

Overall, pasta is more recommendable than rice. As far as its effect on blood sugar is concerned, pasta is considerably better than rice or potatoes. Pasta is formed of a particular netlike protein structure (caution — gluten!) that envelops the carbohydrates, thus slowing down their digestion. Wholemeal pasta is particularly good, although I must admit I have never really grown to love it, despite trying to be as open to it as possible (although, a half-and-half mixture of wholemeal and white pasta tastes great).

Survival factor no. 1: pulses

Among all the high-carb foods, there's one special group that totally outshines all others in terms of health benefits — yet this group is rather unpopular in our culture. I'm talking about pulses — that is, lentils, beans, chickpeas, and peas (botanically speaking, peanuts are also part of this food group, and they are also highly recommended).

Lentils, ugh! I almost never used to eat lentils — nowadays, I eat them a lot, and, each time I put some of the little guys in water to soak, I do a quick check to make sure my wife isn't watching, then give them a bow: that's how great my respect for them is.[35]

All pulses have an extremely low GI (mostly below 50), and they're also high in fibre and provide an excellent source of plant protein — gram for gram, they even provide more protein than a salmon fillet. That's presumably the reason why several studies have found pulses to have a considerable 'slimming' effect: pulses make you *full*.[36]

Thanks to their impressively low GIs, pulses are among the best sources of carbohydrates, especially for people with insulin resistance and diabetes. When diabetics are encouraged to eat more pulses, the proportion of glycated haemoglobin in their blood (the HbA1c value mentioned in the previous chapter) drops within a few months. Their blood pressure, heartrate, and cholesterol levels also fall — which also reduces their risk of developing several diseases associated with old age.[37]

I spent a high-school exchange year staying with a Mexican host family in California, and I've never been treated to so many beans in my life! It's interesting that Mexicans and other Latin Americans living in the US have remarkably lower incidences of chronic diseases, including some kinds of cancer, than the rest of the population, and one of the explanations offered for this phenomenon is their love of beans.

One theory is that the dietary fibre contained in the beans is converted by gut bacteria into fatty acids, which have an anti-inflammatory effect. As some of those fatty acids pass into the bloodstream, they may help to keep inflammation in check throughout the body, thus helping to prevent cancer and other diseases from developing.[38]

A trial at the University of Navarra in Spain confirms this hypothesis, in principle at least. It found that placing test subjects on a diet that included four portions of pulses per week, in the form of beans, peas, chickpeas, and lentils, helped them to lose weight. More importantly, the pulses-rich diet caused a reduction in some inflammatory substances, including one known as CRP, or 'C-reactive protein'.[39] CRP synthesis by the liver increases whenever there's an inflammation somewhere in the body. This protein binds to dead and dying cells and bacteria, which are then consumed by the phagocytes of our immune system. This is very helpful in dealing with acute inflammations. However, a constantly raised level of CRP in the blood is a sign of an immune system on permanent duty, which mostly damages the body it's supposed to be defending.

Since, as described in chapter 2, a permanent, low-level, full-body inflammation is one of the main hallmarks of ageing, pulses are thought to have a positive effect on the ageing process. To test this, an international team of nutrition scientists studied the eating habits of people over the age of 70 in various countries, including Greece, Japan, and Sweden. The question they wanted to answer was whether there is a common denominator in terms of diet and its link to longevity, despite the different culinary traditions.

And the scientists found their answer. As is often the case, their study revealed a positive effect from fish and olive oil. However, the food group that showed up most consistently in connection with an extended life expectancy in all the countries included in the study was pulses. In purely statistical terms, the mortality risk fell by 8 per cent for every 20 grams of pulses — about two tablespoons — eaten per day.

Beyond the bounds of that particular study, it's noticeable that in *every* part of the world where people are unusually long-lived (sometimes called 'blue zones'), pulses are a particularly common food. The Adventists in California eat beans, lentils, or peas every day. On the islands of Okinawa, the traditional diet includes large amounts of soya beans.[40]

Despite the fact that these are once again 'merely' observations, the results of these observations are clearly very consistent. The 'pulses effect' can be seen across cultures, across all types of pulses. In Japan (the country with the highest life expectancy), pulses are popular in the form of tofu, *nattō*, and miso (all soya-based foods). Swedes prefer brown beans and peas. And in the Mediterranean region, the popular pulses are white beans, lentils, and chickpeas, which are eaten, for example, in the form of hummus (GI of hummus = 6. PS: nothing beats home-made hummus — my favourite recipe is in this footnote[41]).

Although all these dishes and foods taste so different, it seems that, once digested, all pulses and their products have basically the same beneficial effect. The research team that carried out the multinational study even concluded that pulses are *the* number one culinary 'survival factor' in old age![42]

Carbohydrates: summary and *Compass* recommendations

There's *no* sound scientific basis for the 'official' recommendation to eat as many carbohydrates as possible. Indeed, it's those carbohydrate bombs that are most popular, such as potatoes and white rice, that can be harmful if consumed in larger amounts — especially with our modern sedentary lifestyle.

This is all the more so for people with insulin resistance. Their bodies do not respond well to a low-fat (therefore high-carb) diet. Those affected by insulin resistance get the most benefit from a low-carb diet — which can be a healthy diet for others, too.

In general, the rule is that it isn't the proportional amount of carbohydrates that's important, but the type of carbs. As far as carbohydrate quality goes, sugar is the low point, especially in the form of an industrially produced liquid such as cola and other soft drinks. When consumed regularly in large quantities, these fructose infusions can lead to fatty liver disease, which in turn can lead to insulin resistance and all the negative consequences that entails, from obesity to all kinds of geriatric diseases. (Remember, two-faced sugar is made up of fructose, which is absorbed by the liver, where it's turned into fat, *and* glucose, which enters the bloodstream, leading in turn to increased secretion of insulin. It's possible that this unique combination of fatty liver disease and increased insulin production is a contributing factor in the overall harmful effects of sugar.)

Pure bombshells of glucose — white bread, white pasta, white rice, and potatoes — aren't quite so bad by comparison, but we simply eat too much of them. They provide large amounts of energy and contain relatively few nutrients. In addition, they cause the levels of sugar in the blood to spike rapidly, though white pasta is digested more slowly and so is less likely to cause sudden spikes in blood sugar and insulin.

Sourdough bread is okay, firstly because its acidity makes it slower to digest. Secondly, sourdough bread is usually made from grain that hasn't had all the nutrients ground out of it, so it contains more vitamins, minerals, and fibre.

In recent years, it's become fashionable to blame bread, wheat, and gluten for all ills. This argument contains a kernel of truth, but goes far too far. It is far more sensible to cut out *processed, rapidly digested, low-fibre* carbohydrates. Wholemeal products are generally to be recommended, including, incidentally, products that we might see as 'exotic', such as bulgur (GI = 48).

Alongside the classic fruit and vegetables, pulses are a particularly valuable source of carbohydrates. Pulses raise blood-sugar levels extremely gradually, they're rich in fibre, and they contain more protein than potatoes, rice, or pasta. This is why pulses also help with weight loss (while potatoes are more likely to have a fattening effect). It's no coincidence that the groups around the world with the highest life expectancy are particularly partial to pulses.

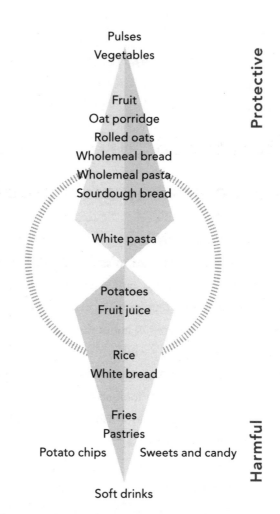

Pulses
Vegetables

Protective

Fruit
Oat porridge
Rolled oats
Wholemeal bread
Wholemeal pasta
Sourdough bread

White pasta

Potatoes
Fruit juice

Rice
White bread

Fries
Pastries
Potato chips   Sweets and candy

Harmful

Soft drinks

Carbohydrate Compass Needle

CHAPTER 7

Intermezzo: drinks — milk, coffee, tea, and alcohol

A co-worker of mine always used to say, 'With the little I eat, I could just as easily drink it.' I'm not quite sure what exactly he meant by that — but when I think of my own appetite for beer, I think I get the general idea. This intermezzo is about the most popular drinks and their impact on our health — milk, coffee, tea, and alcohol (I already dealt with soft drinks and fruit juice in chapter 4). An interesting thing happens when we take a closer look at what we drink: drinks that we generally think of as being good for us turn out to be less healthy than we thought, and vice versa.

Milk

Milk is a complex liquid, scientific findings contradict each other over, it, and it's difficult to come to a definite verdict on it. If you want to hear the short version of my own personal verdict before we go on, here it is: I used to be an enthusiastic milk drinker, and now I'm not.

At first glance, that might surprise you. Despite growing scepticism in recent years, milk still has a very positive image. Furthermore, if you take another look at the diagram in fig. 6.2 (on page 139) showing the results of the large French meta-meta-analysis, you will see that most studies over the course of the decades found milk to be neutral. There are even quite a few in which it comes off positively, and only a very small percentage of past studies came to a negative conclusion about milk. So why am I not a bigger fan of it? The answer is that there are convincing reasons to doubt the reliability of milk's positive image, and to question the accuracy of its 'meta-meta results'.

The first reason for scepticism is the fact that the majority of the studies available to us to make an 'objective' judgement about milk were financed by the milk industry.[1] Of course, there are researchers who are capable of reaching an independent conclusion even if their work is financed by a particular industry or particular interest groups. However, for many, that is demonstrably not the case. The New York nutrition expert Marion Nestle demonstrated this by looking at a sample of 168 industry-funded studies. Of those 168 studies, 156 (93 per cent!) conveniently reached a conclusion that was in the interests of their sponsors.[2] Other investigations have revealed that as soon as a sponsor like the sugar or milk industry has a hand in a study, it is suddenly four to eight times more likely to reach a 'favourable' verdict.[3]

That in itself is far from being proof that milk is unhealthy. It's not poor old milk's fault that it's the product of a powerful industry and that some researchers allow themselves, consciously or unconsciously, to be corrupted. Milk could still be a healthy drink, nonetheless. However, there are now more and more studies, not funded by the milk industry, that suggest this isn't the case (these studies are so new that they were not included in the French meta-meta-study).[4]

Let's begin our critical look at milk with a little context — what we already know about animal protein. Without overstating it too much, we could describe the cow's milk we usually drink as a kind of animal-protein turbo-concentrate. By analogy with fruit juice, it's like giving yourself an intravenous infusion of amino acids. The amino acids shoot into the bloodstream and promptly activate all the growth switches we have met so far: insulin, IGF-1, and mTOR (a basic requirement for the activation of our cellular 'construction manager' is plenty of freely available amino acids in the cell).[5]

In other words, milk is a growth drink. Indeed, it's the ultimate growth drink. That isn't a bad thing in itself: for a baby that needs to grow quickly, breastmilk is certainly the ideal drink. We require no other nourishment for many months of our lives.

The unusual thing about us humans compared to all other animals is that we continue to consume this baby-growing drink into adulthood. But there's another curiosity, too: what we pour into our glass is not human breastmilk, but the milk of another animal species. That's not just unusual, it's also relevant, because cow's milk contains almost three times as much protein as human breastmilk (around 3.4 grams vs 1.2 grams per 100 millilitres), and four times as much calcium. This is one of the reasons why a human baby requires a full 180 days to double his or her weight, while a calf manages it in just 40 days. So when we drink cow's milk as adults, we are consuming an ultra-growth drink during a phase of our lives when we are no longer really growing. As a rule of thumb, an overabundance of growth factors leads to accelerated ageing.[6]

Incidentally, the majority of adults — on a global level — can't tolerate milk at all: their guts are unable to digest the sugar (lactose) contained in milk. We are only able to digest milk as babies because, at that stage of our lives, a gene in our small intestine is activated, enabling the production of an enzyme called 'lactase'. The enzyme lactase breaks down the lactose in the small intestine into its individual parts, which can then be absorbed by the gut. In most people, the lactase gene is switched off in the first few years of life. This is the case, for example, for most people in Asia. The result is that millions and millions of Asian people can tolerate milk only in small amounts, if at all, in adulthood.[7]

Even in a country like Germany, an estimated 15 to 20 per cent of the population are affected by some form of 'lactose intolerance'. Those people can't digest lactose, or can tolerate it only in limited quantities. Since their gut would protest against any milk consumption with flatulence and diarrhoea, they are more or less forced to avoid milk altogether.

The phenomenon of lactose intolerance raises the interesting question of what it means for a person's health. If milk really is essential for us, or somehow crucial to a balanced diet, shouldn't we be worried about the physical wellbeing of all those people with lactose intolerance? Aren't they deficient in protein, calcium, and other important substances? How can a person survive without milk? Since the majority of people clearly do survive, perhaps the better question would be whether living without milk leaves people feeling just a little off-colour, or facing terrible suffering. Does avoiding milk leave people more vulnerable to certain diseases (for example, softening of their bones)?

The answer is no. Quite the opposite, in fact. In some ways, avoiding milk is *better* for us. For instance, people who avoid milk due to lactose intolerance have a lower risk of lung, breast, and ovarian cancer.[8] These findings make sense considering the biological mechanisms involved, not least because cancer growth is stimulated by the aforementioned growth factors, insulin, IGF-1, and mTOR.

Despite all this, we still have to rely mainly on indirect indications when assessing the health implications of milk. It would be more useful if we could establish a definite link between milk and the risk of geriatric disease or mortality. Unfortunately, for the longest time there were simply no reliable scientific investigations into the issue. Until recently.

That's when a team of Swedish researchers set out to get to the bottom of the connection between drinking milk and early death — without funding from the milk industry. The investigation of more than 100,000 Swedes was published in 2017 in *The American Journal of Clinical Nutrition*, one of the most influential nutrition-science publications in the world. Their analysis showed that avid milk drinkers had a 32 per cent *higher* risk of mortality compared to people who don't consume much milk (roughly speaking, they compared people who drank 2.5 glasses of milk a day to those who drank one glass or less per month). But the study also revealed a noticeable exception to this: when consumption of *fermented* milk products was examined, the situation was reversed. People who consumed more yoghurt or cheese were found to live longer![9]

Assuming we are dealing with a causal link here, what might be the reason for this difference? Why should milk be harmful, albeit in high doses, but be beneficial when pre-digested and matured by bacteria? After all, yoghurt and cheese are also known to be rich in animal protein.

Although the scientific results are very consistent in this respect, the mechanism by which the effect is caused remains a mystery. One theory is that the lactic-acid bacteria (lactobacilli) in yoghurt and cheese have a beneficial effect on the gut bacteria in such a way as to check the harmful effects of the proteins and other substances contained in the milk they are made out of.

Another hypothesis, which is both very speculative and very controversial, postulates that the sugar contained in milk, or, more precisely, the galactose, is part of the problem.[10] Lactose is a disaccharide, made up of one molecule of glucose and one of the sugar called 'galactose' (compare granulated sugar, sucrose, which is made up of one glucose and one fructose molecule). The monosaccharide galactose appears to be a molecule that is particularly keen to stick to the protein structures within the body, almost like a kind of biochemical superglue. When they are stuck together, the tissues of the body become increasingly stiff — in other words, the tissue ages.[11] This effect is one of the reasons why scientists use the 'superglue' galactose in ageing experiments with animals. Regular injections of galactose lead to accelerated ageing in mice: the mice develop chronic inflammatory processes, and their brains atrophy, and they die earlier.

Worryingly, this premature ageing process can be triggered in animal experiments by a dose of galactose that would be the equivalent in human terms of just one to two glasses of milk a day (however, this galactose hypothesis can't explain everything on its own, since yoghurt contains at least as much galactose as milk, and only cheese has a lower galactose content).[12]

Whatever the explanation turns out to be, one or two glasses a day seems to be about the critical amount above which, according to current scientific thought, milk becomes harmful for our bodies, or at least for women's bodies. Men may be able to tolerate a little more (three glasses). Research seems to suggest that women have a higher rate of mortality if they drink three glasses or more of milk per day and also eat very little fruit and veg — less than one portion a day. That increase in mortality risk is no less than 179 per cent![13]

Milk: conclusions and recommendations

I didn't set out with the aim of railing against milk — if I had any intention at all, it was the opposite. I was always an avid and copious drinker of milk, and, partially because of the French meta-meta-study, I initially considered all critical findings about milk to be contentious. But when I gradually realised that you can predict with almost 100 per cent accuracy that any positive study was once again funded by the milk industry, my suspicions began to grow, as did my puzzlement, to be honest.

To add to this confusion, recent years have seen the rise of an opposite and very vociferous, sometimes almost militant, anti-milk movement. Its proponents consider it a productive exercise to distort scientific findings extensively to make them appear to support their negative attitude to milk. I've never understood why so many stakeholders and even researchers in the field of nutrition believe such data-bending to be a worthwhile undertaking. Building an objective view is not easy under such circumstances. Nonetheless, I think the following conclusion is fair when it comes to milk.

If you don't like milk — then good for you. Don't drink it. As an adult human being, you do not need milk to be healthy, fit, and happy. And nor do your bones. Of course, it's true that milk contains a lot of calcium, and calcium is good for your bones. But as it turns out, we don't need to consume huge amounts of calcium to build a robust skeleton.[14] The simple fact that far more than a billion Chinese, who are intolerant to milk as adults, don't all end their days in a wheelchair is evidence enough that we don't need to consume milk for our entire lives — an observation also corroborated in non-Asian cultures. What's more, another, also Swedish, study reported a few years ago in the respected publication *The British Medical Journal* found an association between excessive consumption of milk and *more* bone fractures (just by the way, this study also found that the risk of mortality was increased by drinking milk, specifically, while yoghurt and cheese were once again found to be associated with a lower mortality risk).[15] In short, you don't need to drink milk to make your bones strong or to meet your body's requirements for calcium, although, of course, drinking milk can contribute to the latter. The point is that there are healthier sources of calcium than milk, such as yoghurt and cheese, but also wholemeal products and green vegetables, especially kale and broccoli.[16]

If you happen to love milk, my advice is to limit your intake to one or two glasses a day (organic milk from grass-fed or hay-fed cows is the best option). Kefir is a good alternative. I find yoghurt is a very good replacement for milk with your morning muesli. If you're a fan of milkshakes, you can also make them with yoghurt instead of milk. Otherwise: try replacing a glass or two of milk per day with a glass of water, or a cup of coffee or tea.

Coffee

Oh, the irony! Much-praised and often-recommended milk turns out, strictly speaking, not to be a particularly healthy drink for adults. Coffee, on the other hand, is still considered toxic (especially for the heart) by many, despite the fact that drinking it actually lowers the risk of mortality. The ultimate irony may be that one of the ways coffee does this is by protecting against cardiovascular disease.[17] Yes, you read that right: drinking three, four, even five cups of coffee a day is *good* for your heart and the rest of your body, especially your liver. Furthermore, coffee is associated with a reduced risk of various forms of cancer, including not just liver cancer, but also breast and prostate cancer. Drinking coffee also lowers the risk of diabetes and Parkinson's disease by around 30 per cent.[18]

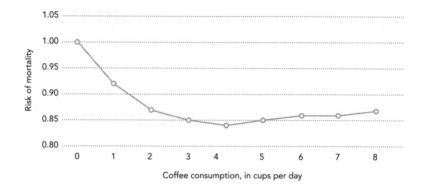

Fig. 7.1 Drinking several cups of coffee a day reduces the risk of mortality by up to 15 per cent — not a huge effect, but also not insignificant. A normal cup contains between 200 and 250 mL of coffee. The graph is based on a large-scale international study using data from almost a million people.[19]

The mechanisms of this beneficial effect of coffee are not very clear, and the mystery is not likely to be solved quickly. The reason is that a cup of freshly brewed coffee is made up of literally hundreds of different substances. Perhaps surprisingly, the healthy effect of coffee cannot, or cannot exclusively, be put down to the caffeine it contains, as decaffeinated coffee has been found to have similar effects.

Some recent new discoveries are of interest in this context. Coffee — with or without caffeine — inhibits the 'construction manager' mTOR and activates a self-cleaning process in our cells, which in turn leads to a real rejuvenation of the cell (more on this in the next chapter).[20] Moderate coffee consumption also seems to keep in check the harmful inflammatory processes that tend to increase with age.[21] Thus, coffee may have an anti-ageing effect on the body via several routes. And if coffee has a beneficial influence on the ageing process per se, it is then no surprise that it also simultaneously protects against so many of the medical problems associated with ageing.

But there is one important qualification to this. The positive health effects mentioned above are mostly only seen with *filter coffee*. This is connected to the bioactive substances coffee contains, which include two oily, fat-like compounds called 'cafestol' and 'kahweol'. Both compounds increase blood levels of 'bad' LDL cholesterol, and blood fats in the form of triglycerides, and both of those are risk factors for heart attacks. In this sense, not all coffee is good for your heart. However, those 'fat molecules' are mostly filtered out by the fine paper of a coffee filter, and filter coffee contains only small amounts of cafestol and kahweol.[22]

It's a different story when it comes to coffee made in the traditional Turkish and Arab way, by boiling the coffee grounds up several times, and even consuming some of the grounds along with the coffee — this creates a beverage that's high in both cafestol and kahweol. The same is true to a lesser extent for coffee brewed in a coffee plunger (cafetière), which works by pressing the coffee grounds down with a steel sieve. Even a small espresso contains relatively large amounts of cafestol and kahweol. An Italian study found concrete evidence of an association between drinking more than two cups of espresso (as opposed to filter coffee) per day and an increased risk of cardiovascular diseases, such as a heart attacks.[23]

In short, drinking three to five cups of filter coffee a day is to be recommended.[24] For espresso lovers, the guideline is: one or a maximum of two cups a day are okay. I would not recommend regularly drinking coffee made from boiling it together with the grounds.

Pregnant women should avoid excessive amounts of not only coffee, but all caffeinated drinks, since caffeine passes easily through the placenta and into the unborn baby's body. This has been associated with, among other things, a lower birth weight (which may — and this is pure speculation — be due to the fact that it inhibits mTOR and therefore also important growth processes).[25] As a rule of thumb, pregnant women should restrict their coffee intake to a maximum of one small cup a day.

Tea

And what about tea? Isn't it healthier than coffee? That claim is often made, but there's a surprisingly limited body of concrete evidence to back it up. Still, since starting my research, I have begun drinking two or three cups of green tea per day, and it's a drink I've really come to appreciate. Overall, I see tea as similar to coffee, although green tea might be a special case; the final proof of that is still to be found. In summary, drinking two to three cups of tea per day reduces the risk of mortality by about 10 (for black tea) and 20 (for green tea) per cent.[26] Pregnant women should exercise caution here since tea also contains caffeine, albeit roughly speaking only about half as much as coffee.

What's the difference between black tea and green tea? It has nothing to do with the species of plant the tea comes from, as they both come from the same one. The difference is in the way the leaves are processed after harvesting. Leaves destined to become green tea are dried rapidly. Such 'pure' green leaves are particularly rich in bioactive substances in the form of polyphenols. Those are substances that, among other things, protect plants from UV radiation or deter herbivores that might otherwise eat them. In the right amounts, those plant protection substances are often very good for our health (more on that in the next chapter).

Black tea is made of exactly the same leaves as green tea. Leaves destined to become black tea are placed, still damp, in a rotating drum — a kind of tea-leaf tumble dryer — where they are thoroughly tossed around. This breaks up the leaves' cell structure. Since this occurs in the presence of oxygen, the leaves become oxidised — the process we know as 'rusting' in the case of iron. One effect of the oxidation process is to make the leaves darker in colour and to lower their polyphenol content. Leaves that start the process green in colour are black by the end of it.

Due to its more highly concentrated polyphenols, green tea is probably (even) healthier than black tea. The most prominent polyphenol in green tea glories in the name of 'epigallocatechin gallate', or EGCG for short (EGCG is present in black tea, too, but in much smaller amounts).

E-G-C-G! The eyes of even the most jaded laboratory researchers light up at the mere mention of those four letters, and with good reason: EGCG can perform all sorts of wonderful tricks in vitro. For example, EGCG can inhibit the growth of all kinds of cancer cells (bladder, stomach, intestinal, liver, lung, skin, and prostate cancer cells, to name just a few). Similarly spectacular cancer-inhibiting properties have also been observed in animal experiments. Incidentally, EGCG also inhibits growth factors like IGF-1[27] and mTOR,[28] and additionally develops a number of beneficial biochemical processes. But that's in the lab. In a Petri dish.

When transferred from the laboratory to the real world, however, the results are a little sobering. Unfortunately, EGCG has not yet been shown to be particularly helpful in combating cancer or any other diseases.

There are very few exceptions to this. One example emerged from a small-scale study in which 60 male volunteers were divided into two groups. All the men had an increased risk of prostate cancer, as determined by biopsy results (some of the cells sampled from their prostate glands appeared under the microscope to be altered, which can in some cases develop into cancer cells). Since it's not practical to brew up a convincing placebo-tea, the men were instead given three pills to take per day. Some were given pills containing a green-tea extract rich in EGCG (equivalent to about six cups of green tea), while the others received pills containing no active substances. After a year, new biopsy samples showed that, as feared, nine of the 30 men in the placebo group had developed cancer. All the men who took the green-tea extract except one remained cancer-free![29]

That sounds promising, but the bad news is that no one has yet managed to reproduce those results. However, there's also some good news. The results of a recent meta-analysis point in a similarly positive direction. They seem to show that drinking seven cups of green tea or more a day actually does reduce the risk of prostate cancer.[30] The main reason I started drinking green tea was EGCG and the euphoric lab results, but I've come to really love the taste, and even love the look of it, with its subtle yellowish-emerald-green shimmer. Now, barely a day goes by when I don't drink any green tea.

Alcohol

I come from a family of wine-growers in the Pfalz, one of Germany's famous wine regions, near the French border. I was practically born among the vines. My father's grandfather was a wine-grower, as were his forefathers, as far back as anyone can trace the family's history. Later, I went to school in Munich and, yes, I'll admit, they do brew one or two good beers there. As I write this, I am in the village near Würzburg, in Bavaria, where I currently live, and where my regular running route takes me through incredibly beautiful wine-producing countryside. In a nutshell, I would like nothing better than to be able to tell you that wine and beer, and alcohol in general, are healthy, in moderation. Can we make that claim, or is it just wishful thinking? Let's take a — hopefully *sober* — look at the current state of scientific knowledge.

There are now *hundreds* of epidemiological studies from all parts of the world that consistently say light to moderate consumption of alcohol lowers the risk of cardiovascular events in particular (incidentally, there is one group who *do not* gain this benefit from drinking alcohol — smokers!).[31] I may come in for some criticism for saying it, but, speaking completely objectively, it appears that total avoidance of alcohol increases the risk of a heart attack by 30 per cent.[32] That figure doesn't come from the German Wine Academy or the German Brewers' Association, which love such findings, but rather is based on a financially independent study carried out in 2017 at the prestigious University College London, taking in data from almost two million people in Britain.

Again, the mechanisms by which this effect might take place are not fully understood, but there are a couple of hot leads. Alcohol causes a rise in 'good' HDL cholesterol, and 'thins' the blood, lowering the risk of a dangerous blood clot (thrombosis). Moderate consumption of alcohol also increases insulin sensitivity and reduces the risk of diabetes — an effect that's rapidly negated by heavy drinking and soon flips to have the opposite effect.[33] And astonishingly, moderate drinking also seems to protect against cognitive degeneration in old age.[34]

It even lowers the overall mortality risk — moderate drinkers live longer than teetotallers.[35] In that sense, not-drinking is a health risk, associated with an earlier death. There's sound data to back this up: women who enjoy up to one 'drink' a day live 1.5 years longer; men who have up to two 'drinks' a day can expect to live 1.3 years longer.[36] Again, these figures stem from a financially independent study carried out by the renowned Karolinska Institute in Stockholm.

So how does that sound? Good, right? That's the kind of data I like. However, there's a little fly in the ointment of these findings, when you look more closely at what exactly a 'drink' is in this context. The study defined a 'drink' as one containing 12 grams of alcohol. That's equal to one 330 mL can of beer, a glass of wine containing 120 mL, or a shot of spirits (40 mL, usually considered a 'double').

A 'drink' might sound vague, but the advantage of this metric is that our culture has trained us to judge intuitively, and with remarkable precision, the alcohol content of a drink by the size of the glass it's typically served in. A typical beer glass contains a similar amount of alcohol to a typical glass of wine or a shot glass (notwithstanding places like Bavaria, where this 'rule' is systematically broken).

By the way, it's easy to work out how much alcohol your favourite beverage contains. As an example, let's take wine with 12.5 per cent alcohol. That means one litre (1,000 mL) of wine contains 125 mL of pure alcohol. To convert that volume into grams, simply multiply the volume by the weight of alcohol (0.8 grams per millilitre): $125 \times 0.8 = 100$ grams. Thus, one litre of wine contains around 100 grams of alcohol. Beer usually contains just under 5 per cent alcohol, which is around 40 grams per litre.

But now for a few sobering facts, just to get them over with. The results of the study can be summarised as such: the absolute optimum, life-prolonging daily amount of alcohol for women is approximately five grams per day; for men, the amount is somewhat larger, say up to twice as much. That means, if you plan to live to the age of 130, you should sip on half a glass of beer or wine per evening if you're a woman, or a whole glass if you're a man. (This is, of course, just the statistically optimum amount; if you drink a few grams more, in the grand scheme of things, you can still expect to see a certain lowering of your mortality risk.)

But that's not all. The positive effect described above only kicks in after we reach a certain age, when our risk of cardiovascular disease — the area where moderate alcohol consumption has its important 'medicinal' effect — starts to reach a significant level, which happens roughly between the ages of 50 and 60. This, too, is not just speculation but the indications from hard data.[37] For anyone younger than 50, there is no positive health effect from drinking alcohol. In many cases, it is simply just harmful, especially for habitual binge-drinkers.

The description above is of the ideal scenario. For most people who like a good drink, that ideal amount will not seem particularly desirable. So let's look at a more realistic question. How much can I drink without ruining my health? Here, the guideline amount is a maximum of two drinks a day for women, and three for men. Anything above that, especially if it's long-term and excessive — including the scenario where moderate regular drinking is interspersed with regular binge-drinking sessions! — increases your risk of all sorts of diseases, in some cases quite drastically.

Excessive alcohol consumption increases in particular the risk of cancers of the oral cavity and throat, as well as the oesophagus. Four drinks or more a day (50 grams of alcohol, i.e. approximately 1.3 litres of beer or 500 mL of wine) increase that risk by no less than 400 per cent, as revealed by the largest meta-analysis on this issue, based on 572 studies.[38] That's an order of magnitude that may be worth remembering the next time you spend the evening enjoying the contents of a whole bottle of riesling while worrying about the pile of fried potatoes you just polished off (a good friend of mine completely fails to see the irony when he enlightens me passionately on occasion about the dangers of eating carbohydrates and other toxic substances, while smoking a post-prandial cigarette).

One reason why the critical amount of alcohol — the amount at which the protective effect flips to a harmful one — is much lower for women is that breast tissue is particularly sensitive to the effects of alcohol and its toxic breakdown product acetaldehyde (it's acetaldehyde that gives us that morning-after hangover). Drinking even small amounts of alcohol is associated with a slightly raised risk of breast cancer.[39]

'Slightly raised' sounds harmless enough, but it becomes significant when the disease in question is common, as is the case with breast cancer. To illustrate this important point, let's take a simple example: suppose a disease affects ten people in every 10,000. Assuming, perhaps over-optimistically, that this book finds 10,000 readers, ten of my readers can then expect to come down with the disease in question. Now let's assume we know that drinking alcohol raises the risk of getting the disease by 10 per cent (which is approximately the actual case for breast cancer). Ten per cent of ten people is one person. 'Only' one extra reader who will get the illness.

Now imagine that the disease is far more common, and affects 500 people in every 10,000. If we then raise the risk by 10 per cent, we are no longer talking about just one extra person getting sick. Ten per cent of 500 is 50, which now means that 50 more of my readers will fall ill!

Since the risk of cardiovascular disease is very high in old age, even a slight reduction in that risk brought about by moderate daily drinking is significant. The same is true of the slight increase in the risk of breast cancer that unfortunately results from even light drinking. There are signs that the increase in the risk of breast cancer brought about by alcohol consumption can be 'cushioned' to some extent by taking extra folic acid, to mitigate the fact that alcohol inhibits the absorption of folic acid in the small intestine. Folic acid is one of the B vitamins and is essential for pregnant women in particular (who shouldn't drink alcohol, but should still ideally begin taking folic acid in tablet form a few months prior to getting pregnant — the generally recommended dose is 400 micrograms a day). Regular alcohol consumption causes the liver to store less folic acid, so more of it is excreted in the urine. In this way, drinking alcohol often leads to folic-acid deficiency. Several studies have shown that women can counteract the rise in the risk of breast cancer from drinking by taking in large amounts of folic acid (women in Germany take in around 230 micrograms per day, which is rather little when compared to their peers in other European countries). Of course, that doesn't give them carte blanche to drink to excess.

By the way, the term 'folic acid' comes from the Latin word *folium*, which means 'leaf'. It got the name because it is found in abundance in green leafy vegetables, including Brussels sprouts, cos/romaine lettuce, and boiled spinach. Other good sources of folic acid are liver, asparagus, lentils, chickpeas, beans, wheatgerm, broccoli, avocados, and oranges.[40]

One final thought on the subject of alcohol: as well as the absolute amount we drink, it's also important *how* we drink it. The speed at which we down it is especially important, as you have no doubt noticed yourself. Our body can't store alcohol, unlike all the other calories we ingest; once it's inside us, our body wants to break it down as quickly as possible. So we can identify some sensible 'rules of play' when it comes to drinking alcohol:

- You can't 'save up' the two to three permissible drinks a day to 'cash them all in' in a weekend-long binge-drinking session (the Russians even have a word for this: *zapoi*). It is far better to drink moderately every day than to run riot for a couple of days in a *zapoi*-style drinking spree.
- Having one or two alcohol-free days a week is a good idea, for at least three reasons. Firstly, to allow your body to 'detoxify'; secondly, because continuous alcohol consumption risks normalising ever-increasing daily amounts; and thirdly, because it makes you appreciate that glass of cool beer or wine all the more when you have it.

- Never drink on an empty stomach, but always as an accompaniment to a meal. And always take your time for both eating and drinking. If you're a wine drinker, always have a glass of water along with your wine. The water is to quench your thirst, the wine is to enhance the aromatic experience of eating your meal.

- It's best to avoid spirits, unless you're the kind of person who can sip on one shot for an entire evening. The typical problem with spirits is that they deliver large amounts of alcohol into the body in far too short a time (and often on an empty stomach).[41] Sugary cocktails are something I would simply never drink.

And here's one more tip: if you're feeling stressed or down, don't turn to the bottle, but rather to your running shoes or your kettlebell, even if it's the last thing you feel like doing at the time. If you're feeling down in the dumps, alcohol will often just make that feeling worse. In vino veritas, as the saying goes: alcohol brings out what's inside us. Running or exercising works off your negative mood. I can almost guarantee it: you don't need to run for more than 40 minutes, and you will feel transformed. Like a new person. (Any exercise that gets your heart pumping and brings you out in a sweat will have the same effect.) And you will maybe even be feeling good enough to enjoy a little glass of your favourite tipple.

CHAPTER 8

Fats I:

an introduction to the world

of fats, using the example of

olive oil

Rapamycin, the anti-ageing agent

Far away from anywhere else, somewhere in the expanses of the South Pacific between Chile and New Zealand, the volcanic island that became known as Easter Island rises from the waves. It's the place that's famous for its mysterious stone sculptures, which the islanders call *moai*. With these colossal heads, several metres in height, they honoured their chieftains and immortalised them. Long-dead yet still massively present, these stone ancestors were a link between this world and the next.[1]

Less famous, but just as interesting, is a bacterium discovered some decades ago in soil samples from that very island. Experiments revealed that the microbe produces a substance with remarkable properties. Scientists named the substance 'rapamycin' — a portmanteau word made up of the first part of the native name of the island, Rapa Nui, and the Greek word *mykes*, fungus, due to the fact that the bacteria uses rapamycin to ward off fungal attacks.

But rapamycin can do a whole lot more, as reported in the scientific journal *Nature* in 2009. No fewer than three research teams working at different labs throughout the US demonstrated, in comprehensive, scientifically sound studies, rapamycin's simple gift of prolonging the lives of mice by up to almost 15 per cent.[2] The findings were particularly remarkable for their consistency. The life-prolonging effect was observed in both females (14 per cent) and males (9 per cent). Rapamycin has this anti-ageing effect on many different genetic strains of mice. Equally promising was the fact that rapamycin even prolonged the lives of mice that were already 600 days old when it was administered. That's equivalent to an age of 60 years in humans. This proves that it's possible to halt a person's biological clock even later in life, and indicates that that may even be a particularly good time to do it. The journal *Science* hailed the discovery of rapamycin as one of the great scientific breakthroughs of 2009.[3]

Further experiments have corroborated the anti-ageing effect of rapamycin. The substance prolongs the lives of *all* the organisms and animals it has so far been tested on, all the way from mice and yeast, to flies and worms. Clearly, rapamycin targets an essential control point of the ageing process. It prevents the development of cancer in mice, but also protects against typical geriatric diseases such as atherosclerosis and Alzheimer's. The fact that it protects against the diseases of old age is a further indication that rapamycin intervenes in the ageing process as such, and slows it down.[4] But how does it do it?

I've mentioned the influential protein molecule mTOR several times already. As a reminder: if there is anything in our cells that can be compared to a construction manager, it's mTOR. When construction materials such as amino acids, as well as energy, are available in abundance, mTOR gives the cell the command to start building, growing, and multiplying (so it's no surprise that mTOR levels are elevated in many types of cancer). If raw materials are in short supply, that activation by mTOR is reduced. In metabolically lean times, cell growth just has to take a back seat. mTOR immediately gives the cell the command to stop all building work.

However, the cell doesn't just wait passively for the crisis to pass and for better times to come. With no other food available, it starts to 'digest' the junk that's accumulated inside it (damaged organelles, agglutinated protein molecules). In a way, cells are no different to us: they become less wasteful and discover the benefits of recycling only when they're in dire straits and nothing else is left.

This clean-up campaign, autophagy, is an important operation. One of the things associated with ageing is an increasing accumulation of molecular junk inside, and sometimes outside, our cells. The waste material hinders the cells in their activity, stopping them from functioning properly and even possibly destroying them — which is probably what happens with Alzheimer's and Parkinson's. To a certain extent, by removing its own rubbish with a self-cleaning program, the cell turns back the biological clock and rejuvenates itself.[5]

This is the point at which rapamycin comes into play. Rapamycin inhibits mTOR, hence its name, which is an abbreviation of 'mechanistic target of rapamycin' — it's the docking target for rapamycin.[6] The most effective way to extend an animal's life span, as we know, is to put them on a permanent starvation diet. Calorific restriction reduces the activity of mTOR and provokes the cell's appetite for autophagy. A more pleasant alternative to calorific restriction is to eat as much as you like, then take rapamycin. Rapamycin enters the cells of the body, attaches to mTOR, and deactivates it, even when there's no shortage of nourishment. It's as if rapamycin dupes the cells into thinking famine has struck; construction is halted, and autophagy begins.

In other words, rapamycin sounds like the perfect anti-ageing agent — if it weren't for a few less-than-pleasant side effects, such as a strong suppression of the immune system, insulin resistance, cataracts (clouding of the lens of the eye), and, in men, testicular degeneration, I'm afraid.[7] Some people might dismiss these risks and side effects as 'peanuts' — and in the case of testicles, they would be more correct than they might like — if the reward is a longer life free of cancer and Alzheimer's. But the fact is that no one knows the full effects of rapamycin, or in what dosage they become significant.

Nevertheless, there are indications that we may be able to gently inhibit mTOR without the side effects. And in a completely natural way. I think you might already have an idea how that could be done. That's right, with our diet. Which brings me to our third and final major food group: fats.

As I've already said, one of the most surprising things I learned while researching *The Diet Compass* was that there's no actual basis for our widespread fear of fat. The fat we eat does not necessarily make us fat, nor is fat fundamentally harmful. A high-fat diet helps people with insulin resistance to lose weight better than a classic low-fat diet. In addition, many fatty foods are purely and simply good for us — often being far healthier than the rapid carbs we replace them with, such as potatoes, rice, and white bread. I can't stress often enough that this is not meant to be a controversial statement, but simply a clearheaded summary of the scientific evidence that has built up over the decades. By the way, I myself now eat more fat than I used to (and I'm slimmer and feel better).

The 'ageing switch' mTOR could be central to all this. I've already covered the fact that, of the three main food types, it's proteins (amino acids) that really get mTOR going. The second crucial mTOR activator is glucose together with insulin — which would indicate that carbohydrate-rich foods with a high glycaemic index, like potatoes, rice, and white bread, are unhealthy because they boost the ageing process. Of course, fat also forms part of our energy supply, which is registered by mTOR. As a rough rule of thumb, however, we can say that, among the three main food groups, firstly, carbohydrates with a low GI ('slow carbs', for example in the form of pulses) and, secondly and just as importantly, fat are among the substances that leave mTOR relatively 'in peace'.[8]

Whether it's via that route, or another, or some combination of the two, many foods with a high fat content are extremely good for us. We should be eating more of them. Some examples of such foods are: olive oil, nuts, avocados, and even dark chocolate, which is a whopping 50 per cent fat (cacao butter). And then there are the particularly beneficial omega-3 fats, which are found in wholemeal products, chia seeds, linseeds, walnuts, and rapeseed oil. The main source is, as I'm sure you've often heard, oily fish such as salmon, herring, mackerel, sardines, and trout. And there are other healthy fats, called omega-6 fats, which are found, for example, in sunflower seeds and sunflower oil.

So, fat. The subject of the next three chapters. I will be looking at wonderful, sometimes extremely healthy fatty acids and fatty foods. I'm afraid your mouth will water when you read about them. I wouldn't be surprised if you have to stop reading from time to time to pop to the kitchen with a rumbling stomach. Whatever the case, I'm sure that when you reach the end of these three chapters on fats, you'll have been cured of any fatphobia you may have had. You'll enjoy eating fat more than ever before, and so you should.

Olive oil: heart killer or liquid gold?

Where would Mediterranean food — and indeed, food in general — be without olive oil? Olive oil, described as 'liquid gold' by the Ancient Greek poet Homer, is not just kind to the tongue and the palate. This oil could very well turn out to be a delicious medicine for the whole body. It was only recently, for instance, that the researchers who carried out the Spanish study into the Mediterranean diet (see chapter 3) made an astonishing discovery while they were reviewing their results again. They noticed that the women who were lucky enough to be placed in the olive-oil group — who each received a free litre of high-quality olive oil every week — had a 68 per cent lower risk of getting breast cancer than those in the control group. Up to a certain point, there was even a clearly recognisable dose-response relationship: the more olive oil a woman consumed per day, the lower her risk of breast cancer was.[9] Since, unfortunately, breast cancer is a relatively common disease, this finding is highly relevant.

That sounds promising. Yet not wanting to chime in prematurely with Homer's hymn of praise for the oil, I should first examine a widespread fear that fat, even the fat in olive oil — despite its ability to protect against certain diseases such as breast cancer — ultimately helps to 'block' our arteries (just as the drain in your kitchen sink gets blocked if you pour too much fat down it). There are low-fat advocates that certainly argue this way, usually hard-core vegans. Their line of argumentation, which is certainly plausible, goes like this: Yes, the Mediterranean diet is healthy. But the reason for that is that much of the Mediterranean diet is made up of fruit and vegetables, pulses, and wholemeal products, and it has nothing at all to do with olive oil. Quite the opposite, in fact, and if olive oil is removed from an otherwise Mediterranean-type diet, it becomes even healthier!

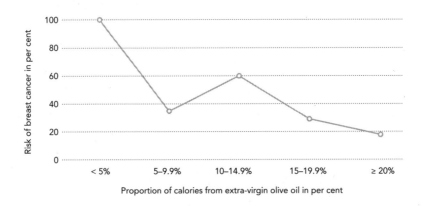

Fig. 8.1 The higher the consumption of olive oil, the lower the risk of breast cancer. Women who love olive oil (20 per cent or more of their daily calorific intake) have a risk of breast cancer that is around 80 per cent lower than that of the women who consumed the least olive oil (less than 5 per cent of their calorific intake).[10]

A prominent proponent of this position is the US cardiologist Caldwell Esselstyn. For years, he has been treating a small group of heart patients whose hopeless situation meant their previous cardiologists had written them off. Esselstyn places his patients (who call him 'Essy') on a radically low-fat, vegan diet. That means no meat, no animal products at all, no milk, no eggs, no butter, no cheese, no honey. Esselstyn and his disciples eat an entirely plant-based diet of wholemeal products, vegetables, pulses, and fruit. Esselstyn advises against eating nuts and avocados as they are too fatty. But his main maxim is — no oil. 'Not even a drop!' as Dr Esselstyn advises in his lectures as well as his very worthwhile book *Prevent and Reverse Heart Disease*, because, '[each spoonful] is every bit as aggressive in promoting heart disease as the saturated fat in roast beef'.[11]

I have a lot of respect for Dr Esselstyn. His uncompromising nutritional concept is certainly no picnic, but for those who manage to keep to the 'Essy' diet, it seems to work wonders. Although Esselstyn has meticulously detailed the beneficial effects of his heart-protecting diet, what he is doing is not science in the strictest sense (for example, he has no control group, which is one of the reasons why most researchers ignore him, although there are a few scientifically rigorous experiments that seem to support Esselstyn's results, or at least seem to point in the same direction).[12]

Some of his patients were, as Esselstyn puts it, closer to death than to life.[13] And yet, just a few weeks to months after changing their diet, almost all of them felt like a new person. Many of them were now able to walk without feeling a pain in their chest or experiencing breathing difficulties, and were even able to exercise. X-rays revealed that some patients' massively damaged blood vessels were spectacularly repaired (see fig. 0.2 in the introduction).

Having suffered myself from similar symptoms, I also changed my diet — initially as an experiment, then permanently — to make it mostly plant based. I became like a long-distance 'Essy' patient. For the first time in my life, my plate predominately contained mixed salad with spinach leaves, broccoli, carrots, zucchinis (courgettes), onions, Brussels sprouts, beans, or my beloved lentils. I still try to eat as many vegetables as I can, which I sometimes manage better than other times.

Of course, I've often wondered what the decisive factor in my recovery was (it was probably everything together). Whatever the case may be, I can say without reservation that it worked, and remarkably quickly. I was already feeling better three or four weeks after I changed my diet, which I first noticed when I 'stress tested' my body by going out for a run again — and those attacks of pain in my heart disappeared astonishingly quickly. It then took a few more months, a full year overall, before my last heart 'stumbles' were completely gone. And since then, my heart problems have disappeared. And I mean *completely* disappeared. One of the things that left a lasting impression is the fact that I *never again* had one of those night-time attacks. Something inside me had changed fundamentally, and for the better.

One thing's clear about my recovery: not eating fats can't have played a part in it because, as I've already described, I now eat far more fat than before, albeit now (almost) always healthy fat. In particular, I eat far more nuts, olive oil, and natural peanut butter;[14] I have avocados every week; and I eat more oily fish, more chia seeds and linseeds, and more dark chocolate. (For a time, I experimented systematically with eating less fat and noticed no difference as far as my symptoms were concerned.)

This personal experience and, more importantly for the purposes of this book, the mass of scientific evidence, convinced me that the reason Esselstyn's dietary approach is so good for the heart is not *because* he bans all fats, but *despite* it. The evidence for the beneficial effect of eating nuts is particularly impressive. Esselstyn is definitely wrong about nuts. You should eat a handful every day, and they needn't just be walnuts. Even peanuts (which aren't really nuts, botanically speaking, but pulses) are highly recommended.[15]

Avocados are also a good recommendation, and I find Esselstyn's rejection of them strange and pointless. There's good evidence that eating one avocado a day has a positive effect on blood-fat levels, which reduces the risk of cardiovascular disease.[16]

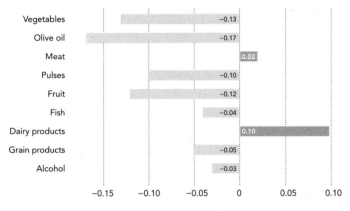

Fig. 8.2 How closely is each component of a Mediterranean diet linked to the risk of cardiovascular disease? I find this recent analysis to be a good summary of some of the most important components of that diet and their link to healthy eating: vegetables, pulses, and fruit form the basis — and yes, in this context it's fine to use generous amounts of high-quality olive oil. After researching this book, I am no longer a big fan of milk, and it makes sense to differentiate between different dairy products. Yoghurt, for example, is to be recommended, cheese is also more positive than negative, and butter can be considered 'neutral'.[17]

The vast majority of studies also support the theory that olive oil is beneficial, especially, but not only, for the heart. A recent analysis by an international team of researchers agrees. Their study, which analysed the findings of just under a dozen observational and experimental studies, yielded some revealing results. When the individual components of the Mediterranean diet are analysed, it turns out, as expected, that many of them have beneficial health effects and lower the risk of cardiovascular disease. Some, by contrast, actually result in a slightly *higher* risk of heart problems (see the bar chart in fig. 8.2). I'll give you three guesses as to which food is associated with the largest *fall* in the risk of cardiovascular disease (it's olive oil). Of all the different components making up the Mediterranean diet, it is the fattiest that provides the most protection for the heart![18]

A short ABC of fatty acids

What makes olive oil so special? To answer that question, I need to take you a little deeper into the world of fats.

For the most part, olive oil is made up of a fat called oleic acid, a monounsaturated fatty acid. The fat molecules in the food we eat are mostly stored in the body in a form whose name we have met before: triglycerides. A triglyceride is made up of three ('tri-') fatty acids, held together by a kind of bracket ('glycerol'). You can imagine a triglyceride as being like a fork with three prongs. As it's the fatty acids — the individual prongs — that are important for health, I will spend a few pages describing them in more detail.

Put simply, a fatty acid is a chain of no fewer than two and no more than 30 carbon atoms (C), mostly with two hydrogen atoms (H) attached to each of them. Here's an example:

$$\underset{\text{HO}}{\overset{\text{O}}{\diagdown}}\text{C}-\underset{\text{H}}{\overset{\text{H}}{\text{C}}}-\underset{\text{H}}{\overset{\text{H}}{\text{C}}}-\underset{\text{H}}{\overset{\text{H}}{\text{C}}}-\underset{\text{H}}{\overset{\text{H}}{\text{C}}}-\underset{\text{H}}{\overset{\text{H}}{\text{C}}}-\underset{\text{H}}{\overset{\text{H}}{\text{C}}}-\underset{\text{H}}{\overset{\text{H}}{\text{C}}}-\underset{\text{H}}{\overset{\text{H}}{\text{C}}}-\underset{\text{H}}{\overset{\text{H}}{\text{C}}}-\underset{\text{H}}{\overset{\text{H}}{\text{C}}}-\underset{\text{H}}{\overset{\text{H}}{\text{C}}}-\underset{\text{H}}{\overset{\text{H}}{\text{C}}}-\underset{\text{H}}{\overset{\text{H}}{\text{C}}}-\underset{\text{H}}{\overset{\text{H}}{\text{C}}}-\underset{\text{H}}{\overset{\text{H}}{\text{C}}}-\text{H}$$

As you see, there are two hydrogen atoms 'stuck' to each carbon atom (only the ends of the chain have a different configuration, but that's not important for us here). The fatty acid is, as we say, *saturated* with hydrogen atoms — hence the name 'saturated fatty acid'.

Food always contains a mixture of different fatty acids, although one form often dominates. Other common sources of saturated fatty acids include full-fat milk, red meat, and cheese. There's no reason to demonise saturated fatty acids, but, overall, they shouldn't be eaten to excess by people who've been diagnosed by their doctor with high blood-cholesterol levels. Most saturated fatty acids raise the levels of 'bad' LDL cholesterol in the blood.

Saturated fatty acids with mid-length carbon chains (six to ten carbon atoms) are a positive exception in several ways, known scientifically as MCT, or 'medium-chain triglycerides'. There are modest amounts of MCTs in coconut oil, cheese, milk, and yoghurt. They are only contained in highly concentrated form in the MCT oil mentioned in chapter 5. It boosts fat burning and helps with weight loss. MCTs also increase insulin sensitivity.[19]

Thanks to its saturated state, the carbon chain of a saturated fatty acid is as straight as a toothpick, which means that several of them can be packed closely together, in the way many toothpicks can be kept in a small container, each taking up very little space. This is why saturated fatty acids are usually solid at room temperature — like butter, for example.

Olive oil also contains a little over 10 per cent saturated fatty acids, but most of the fat it contains, usually a little over 70 per cent, is made up of *monounsaturated* fatty acids. Other sources of monounsaturated fatty acids include avocados and poultry meat, as well as many nuts, including macadamias, hazelnuts, pecans, almonds, cashews, and peanuts. About half the fat contained in red meat is in the form of monounsaturated fatty acids (the other half is saturated fat). The molecular structure of monounsaturated fatty acids looks like this:

Two hydrogen atoms are missing at a precise position in the carbon chain. At that position, the carbon atoms form what chemists call double bonds (C=C). Since both missing hydrogen atoms are on the same side of the chain, there's a gap, which causes a bend in the molecule. The fatty acid looks like a broken toothpick, and the molecules can no longer be packed so closely together. This is the chemical reason why oils such as olive oil are liquids. And it turns out that what makes for bad toothpicks, makes for good health benefits. The bend in the molecules of unsaturated fatty acids means they are stored in a more 'aerated' way. This has beneficial consequences for the body, which itself consists to a great extent of fatty acids.

Our cell membranes, for example, are made up of fatty acids. The brain is a particularly fatty organ, and some of the fat it's made up of comes from the fatty acids we eat. If we eat mainly saturated fats, our cell membranes become stiffer, resulting in, if you like, a 'stiffer' brain (and from there, it's just a short hop to being a bonehead). Eating more unsaturated fatty acids, such as olive oil and omega-3 fats, leads to more supple cell membranes. That suppleness is important because our cell membranes are extremely dynamic structures. They are spotted with countless receptor molecules and channels that pass through the fatty membrane (scientists call these 'lipid rafts'). Some of these 'floating' structures act like antennae, passing signals from outside the cell to its interior. In this way, glucose, vitamins, and other nutrients find their way into the cell. When our cell membranes are more flexible due to a greater proportion of unsaturated fats in our diet, those substances are able to pass through the cell membrane more easily. And that means our cells can simply carry out their functions better.

So this gives us our first rule of thumb: unsaturated fatty acids are healthier than saturated fatty acids. This really is nothing new, and it's a confirmation of the basic assumption that underpins the low-fat argument, which has always recommended avoiding saturated fats. But let's not get carried away: saturated fatty acids are not so bad that we need to ban them from our diets. They can be perfectly fine, and even healthy, especially in the form of cheese or MCT oils (more on saturated fatty acids in the form of butter and cheese in the next chapter).[20]

However, there are some kinds of fat you should steer well clear of. The main one is the Frankenstein fat called 'trans fat' by nutrition scientists. Trans fats, at least the harmful ones, are an industrial product. They are the result of the artificial hardening of unsaturated fatty acids, which are otherwise found in liquid form as oils, so that they turn into spreadable margarine by the time they leave the production line. Take a look at the remarkable structure of trans fats:

$$O \overset{\displaystyle \,}{\underset{HO}{\diagdown}} C - \overset{H}{\underset{H}{\overset{|}{\underset{|}{C}}}} - \overset{H}{\underset{H}{\overset{|}{\underset{|}{C}}}} - \overset{H}{\underset{H}{\overset{|}{\underset{|}{C}}}} - \overset{H}{\underset{H}{\overset{|}{\underset{|}{C}}}} - \overset{H}{\underset{H}{\overset{|}{\underset{|}{C}}}} - \overset{H}{\underset{H}{\overset{|}{\underset{|}{C}}}} - \overset{H}{\underset{H}{\overset{|}{\underset{|}{C}}}} - \overset{\mathbf{H}}{\overset{|}{C}} = \overset{}{\underset{\mathbf{H}}{\underset{|}{C}}} - \overset{H}{\underset{H}{\overset{|}{\underset{|}{C}}}} - \overset{H}{\underset{H}{\overset{|}{\underset{|}{C}}}} - \overset{H}{\underset{H}{\overset{|}{\underset{|}{C}}}} - \overset{H}{\underset{H}{\overset{|}{\underset{|}{C}}}} - \overset{H}{\underset{H}{\overset{|}{\underset{|}{C}}}} - \overset{H}{\underset{H}{\overset{|}{\underset{|}{C}}}} - H$$

A trans fat is like a broken toothpick that has been clumsily bent back into shape. Trans fats are unsaturated insofar as they are also missing a few hydrogen atoms, only not in the usual position on the same side of the molecule. Instead, one hydrogen atom has been transferred to the other side of the molecule, hence the name 'trans fats'. That makes the gap smaller, and the bend in the chain is largely gone.

Trans fats have turned out to be downright toxic. It's not just that they stiffen our cell membranes, they also have the worst impact possible on our blood-fat levels. Trans fats increase the amounts of the 'bad' LDL cholesterol and triglycerides in our blood, while simultaneously lowering the amount of 'good' HDL cholesterol — quite a feat for a fat. Trans fats increase in particular those dangerous little LDL particles (small, dense LDL, or sdLDL, covered in detail in chapter 4). And as if that weren't enough, trans fats also promote inflammation and lead to insulin resistance. All this means it should come as no surprise that one of the health effects of trans fats is to increase massively the risk of cardiovascular disease.[21]

In addition, trans fats are among the very few fats that really do make us fat. In one experiment, scientists fed two groups of monkeys an almost identical diet for six years. The only difference was that one group's food included monounsaturated fats, while the other group's dietary fat was presented in the form of trans fats. The monkeys were neither starved nor intentionally fattened. The number of calories they were given was meticulously calculated on the basis of their body mass to keep their weight as stable as possible (they received 70 calories per kilogram of body weight per day).

After six years, the results showed that the monkeys in the group that received the monounsaturated fats had maintained a stable body weight, as expected. The trans-fat monkeys, by contrast, had gained almost half a kilo in weight — although they received the same, carefully calculated, number of calories! Half a kilo? You may be thinking that's not so bad. But don't forget these monkeys only weigh about seven kilos in all. Their weight gain would be the equivalent of a 70-kilo human putting on 5 kilos! But at least as significant was the fact that the extra fat had mainly accumulated in the monkeys' abdominal region, and the animals showed clear signs of insulin resistance.[22] In short, trans fats make you fat *and* ill. They should be avoided at all costs.

Although industry has reacted to this news, and trans fats are now gradually disappearing from food processing, they can still be found in fries, potato chips, and other (deep-fried) fast food, ready-made pizzas, donuts, cookies, biscuits, and other industrial baked goods. Some margarines still contain trans fats. The US, Canada, and some other countries have banned artificial trans fats completely — Germany is not among them; nor are Britain, Australia, and New Zealand. Unfortunately, in the absence of a ban, there's rarely even a duty of declaration, so there is no way for consumers to tell how much trans fat is in the food they buy. This is why I, personally, now never eat fries or any baked goods I didn't make myself. At the bakery, I ignore the front display cabinet — the fat and sugar zone that attracts so many wasps and bees in summer (insects not exactly known for their longevity — and no wonder, with that diet …).

Finally, there are also *polyunsaturated* fatty acids, which include, firstly, omega-3 fats, found mostly in oily fish, as mentioned before. The second variety is the omega-6 fatty acids, to be found in many types of nuts, seeds, and oils such as sunflower oil and some types of thistle (safflower) oil. The exotic-sounding names 'omega-3' and 'omega-6' simply give information about the position in the carbon chain of the first double bond, i.e. where the first kink is, counting from the end of the molecule (for those whose Ancient Greek is no longer as fresh as it might be, here's a little reminder: omega is the final letter of the Greek alphabet, hence the phrase 'the alpha and the omega'). In omega-3 fatty acids, the first C=C double bond is located on the third-to-last carbon atom, as seen here:

'Polyunsaturated' means these chains have several kinks, which give our cell membranes the flexibility of an Indian yogi — surely that's the ultimate high-point of healthy food ... That may be a great oversimplification, but there is a kernel of truth in it. And it's underpinned by scientific findings from recent years, including from research at Harvard University. That particular study followed the fates of more than 126,000 people over a period of 32 years. The central aim was to find out how mortality risk is affected by replacing some of the carbohydrates in a person's diet with a corresponding amount of various fats. The graph in fig. 8.3 shows a summary of the results from that study. In very general terms (after all, it ultimately comes down to individual foodstuffs), we can conclude that replacing carbohydrates with saturated fats increases the risk of mortality. Replacing them with unsaturated fats has the opposite effect. Polyunsaturated fats turn out to be the healthiest option of all.[23] (A new study published in the medical journal *The Lancet*, involving data from more than 135,000 people from 18 countries, generally comes to a similar conclusion, although, in this study, saturated fatty acids also come off relatively well.)[24]

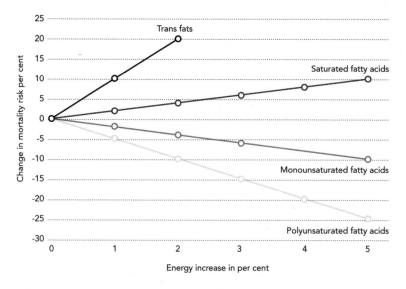

Fig. 8.3 The graph summarises the change in mortality rate associated with replacing dietary carbohydrates with different kinds of fat. Trans fats (for example, in the form of fries and industrial baked goods) drastically increase the risk of mortality. Polyunsaturated fatty acids (found in fish, walnuts, and such oils as sunflower oil and rapeseed oil) reduce that risk.[25]

If it makes you cough twice, you've found a good oil

The yoga-factor of fatty acids is not the *only* measure of the healthiness of high-fat foods. As always, the entire package is what's important. Once again, olive oil is a good example. The fact is that even olive oil is made up of more than just fat molecules. The original olives it's pressed from are teeming with so-called phytochemicals. What are they? Well, life isn't so easy if you're an olive. If the sun is scorching, you can't just move into the shade. You can't just jump under a cool shower when the weather's too hot. When fungal infections threaten, you can't just run away. But necessity is famously the mother of invention, and the olive has developed chemical defences against such threats, and has an entire arsenal of chemical weapons, the phytochemicals.

The chemical structure of those substances means they are classified as 'polyphenols'. Two such polyphenols are known as 'oleuropein' and 'oleocanthal'. That might sound like chemical gobbledygook, but you might be surprised to know that you can taste oleuropein and oleocanthal. Oleuropein tastes bitter, and oleocanthal has a sharp, peppery taste, creating a scratchy feeling in the throat when you taste a tablespoon of good olive oil.[26] Many polyphenols, including oleuropein and oleocanthal, have an almost medical effect on our bodies. A *Nature* study hit the headlines a few years ago after a researcher had a light-bulb moment, realising that the scratchy-throat feeling caused by a solution of the painkiller ibuprofen is very similar to that caused by good-quality olive oil (if you've ever let an ibuprofen tablet dissolve in your mouth, you'll know the feeling).

The resulting studies revealed that oleocanthal inhibits the same inflammatory signal pathways as ibuprofen, albeit in practice to a much lesser extent since the dose is so much lower. Fifty grams of cold-pressed olive oil would have the equivalent effect to a normal ibuprofen pill. This watered-down effect is not necessarily a bad thing, however. In fact, one of the reasons olive oil is so good for us may be that oleocanthal reduces the chronic inflammation typical of old age in a *gentle* way.[27]

It's possible that the polyphenols in olive oil also have a much more direct anti-ageing effect on us. Recent experiments have shown that both oleocanthal[28] and oleuropein[29] are inhibitors of our old friend mTOR — olive oil may turn out to be a kind of tasty rapamycin with a rejuvenating effect on our cells!

That's pure speculation at the moment, but it's already pretty clear that the phytochemicals that plants use to ward off all kinds of attacks are among the healthiest substances nature has to offer.[30] In view of the aggressive solar radiation the average olive is exposed to day after day, it may come as no surprise that consuming olive oil — according to some thoroughly sound studies — is also associated with slower ageing of the skin caused by UV radiation.[31]

Since studying the research results so closely, I have learned to appreciate stressed and wrinkly-looking vegetables far more than the perfect, shiny, showroom veg that upmarket supermarkets pride themselves on so much (just look at the difference between a showroom lemon and a real, organically grown lemon, which has experienced all the ups and downs of a natural lemon life). A child who grows up spoilt will not develop resistance to stress. We may appear to be doing those delicate little plants a favour by mollycoddling them and providing them with the most perfect conditions possible to grow in (nice and warm, always well watered, an uplifting psychodynamic pep talk every day …). And this pampering program is never complete, of course, without regular spraying the plants with two dozen different pesticides. After all, a microbial attack would put far too much strain on our delicate babies! But this all-round wellness program also removes the need for the plants to develop stress-busting polyphenols to defend themselves.[32]

In short, olive oil is far more than just a concentrated shot of fat. Allow yourself the luxury of good, cold-pressed, extra-virgin olive oil — a particularly virginal virgin, as the Italians say. Good oils taste bitter and peppery, the flavours of oleuropein and oleocanthal. It's no accident that professional chefs often say of olive oil that 'the quality is in the cough'. If tasting a pure oil makes you cough, or, even better, cough twice, you can be sure you have a decent one. In industrially processed ('refined') oils, the phytochemicals have mostly been decimated, and I advise against their use.

Unfortunately, even among connoisseurs of olive oil, the stubborn myth persists — possibly introduced by industrial manufacturers as a way of selling more lower-quality oil — that extra-virgin olive oil should not be used for frying. The myth goes that simple, 'refined' oil is better for this purpose. Wrong! The results of studies of frying are unambiguous in showing not only that olive oil is a fundamentally very stable oil, even at high temperatures,[33] but that the polyphenols contained in good-quality oil can even help prevent the development of carcinogenic substances when red meat is fried.[34]

I use extra-virgin olive oil at home almost exclusively. For dishes that require a more neutral-tasting oil, my wife prefers a cold-pressed organic sunflower oil, which is also a good oil if used in moderate amounts, and it's especially good the way my wife uses it. However, for those who simply don't like olive oil, I would actually recommend a cold-pressed rapeseed oil, which has a similar fatty-acid profile to olive oil but with far more plant-based omega-3 fats. And, I find, it also tastes excellent.[35]

So now I've sung the praises of olive oil enough! I will say not another word about it. Please do not consider this exposition as some kind of advertising campaign for the olive-oil industry, but simply as an introduction to the subject of healthy fats, using the best example possible.

CHAPTER 9

Fats II: saturated fatty acids — palm oil, butter, and cheese

Only some fats make you fat; others make your muscles

Imagine I'm your gran. I want to fatten you up, so I give you three muffins every day. You eat them with gusto, in addition to your usual diet. You're now eating 750 calories more each day than usual.

Being a kind-hearted gran, I also give muffins to your partner or your best friend. After a couple of weeks of being spoiled by me, you will both have gained some weight. To make sure you're both putting on the same amount of weight (I'm a very fair-minded grandmother), I put you on the scales regularly.

That's it. Oh no, there is one other thing. A small detail. Just for fun, I bake *your* muffins using polyunsaturated fatty acids (sunflower oil). Your partner gets the same muffins, with one small difference: instead of using unsaturated fats, I use saturated fats (palm oil). The muffins are otherwise completely identical, including their calorie content, of course.

What do you think will happen? Will some difference between you and your partner be revealed in the end? Since your muffins have the same calories, and if I also take special care to make sure you both put on the same amount of weight, there shouldn't be any difference between you …

Initially, that appeared to be the case when researchers at the University of Uppsala in Sweden carried out this 'grandmother experiment'.[1] The test subjects in both groups were fed the muffins for seven weeks, and people in both gained an average of 1.6 kilos. Exactly the same amount. From the point of view of the scales, the two types of muffins had the same effect on the test subjects' bodies. If the researchers had ended their experiment there, they would have undoubtedly reached the usual conclusion: if you eat more calories than you burn, you will inevitably gain weight, full stop. A calorie is always a calorie.

But the Swedish scientists took their experiment a step further and used magnetic-resonance imaging to take a close look inside their guinea pigs' bodies. And they found some very revealing differences. In the sunflower-oil group, only about half of the weight gain was due to the formation of fresh fatty tissue. The other half of their weight gain did not appear to be fatty tissue at all — the extra calories had been turned into 'lean' *muscle tissue*! Interim conclusion: some fats contribute not only to fat-building, but also to muscle-building, even when we overeat. That, by itself, is an astonishing finding.

What's even more remarkable, however, is the contrast to the palm-oil group. There can certainly be no talk of muscle building in that group. On the contrary, not only did those people's livers became fatty, but the subjects gained a lot more abdominal fat. Thus, saturated fats contribute to (harmful) fattening of the body, confirming what most of us think.

Yet on even closer examination, the results of this study are not quite so unambiguous. But first things first: there's no doubt that extra calories cause extra body weight. Calories can't just disappear into thin air. All the test subjects had gained weight by the end of the experiment. So we can confirm that eating calories without burning them leads to weight gain. But the question is, what do our bodies do with those additional calories? Where do they go? How are they 'distributed' throughout the body? Clearly, that depends on what kind of food the calories come from.

Before we fall back into the habit of demonising saturated fats, we must remember that the muffins weren't made up *only* of fat. They had to be edible, so both sets also contained a lot of fructose. That means it might have been chiefly the fructose that the subjects were eating that caused the ones in the palm-oil group to gain liver and abdominal fat. We know that fructose has the ability to do that to people, in principle at least. But we can't dismiss the possibility that the palm oil itself contributed at least a little to making the test subjects internally fatter. It's also possible that the combination of fructose and palm oil is particularly fattening.

Whatever the reason for the increase in liver and abdominal fat, the contrast with the subjects from the sunflower-oil group was astonishing. *Although* the subjects in that group also ate three muffins heavily laced with fructose every day (which works out at about 150 muffins in total, in addition to their normal diet), they showed no increase in liver fat at all. How can we explain that? It must be linked to the polyunsaturated fats. The probable explanation is that polyunsaturated fatty acids can actually protect the liver from becoming fatty — even during a fructose attack of several weeks' duration. In fact, there are indications that polyunsaturated fatty acids can simply *switch off* certain genes that promote the formation of fat in the liver. That means some fats not only *don't* make you fat, they can even have a 'cushioning effect' against too much fattening of the body, even when we are constantly overeating.

To sum up: the Swedish muffin experiment is another piece of evidence for the idea that unsaturated fats have a more positive impact on our bodies than saturated fats — even though the number of calories they provide is the same. And even for those who eat a very high-calorie diet, polyunsaturated fats can offer protection against the accumulation of internal fat.

There's another important detail concerning saturated fatty acids: the Swedish scientists chose palm oil as the source of saturated fats in their experiment — and palm oil does not have a very good reputation, although the scientific evidence for one side or the other is sparse and rather contradictory.[2] Palm oil is derived from the oil-palm plant and is particularly beloved of the food industry because it's cheap and flavourless. That's why it is used in many industrial products, especially margarines, but also creamy spreads (Nutella, some peanut butter), ice cream, cookies and other baked goods, ready-to-eat pizzas, and sometimes even sausages.[3] Since these foods aren't exactly healthy fruit and vegetables, avoiding palm oil seems like a sensible idea in general to me. The Swedish muffin study doesn't do the oil's reputation any favours, either. I advise caution here until more results come in.

The situation concerning margarine is similar. The problem with margarines is not just that they may contain trans fats and palm oil, but also that it's often difficult to know just what is in some of them. As far as trans fats are concerned, the situation has improved enormously — however, I advise avoiding all sunflower margarine, which is infamous for containing trans fats. Sunflower *margarine* is not the same as sunflower *oil!*[4] To be sure, I avoid margarine altogether, even if some kinds may turn out to be okay in the end.

Butter: healthier than the white bread we spread it on?

In another, not quite so well-controlled study, the same Swedish research group showed that there is a similar effect to the palm-oil-liver-fat effect when the palm oil in the muffins is replaced with regular butter.[5] As we know, butter consists primarily of saturated fats. This led the scientists at Uppsala University to suspect that the negative effects were of a general nature and could be expected from all, or at least most, saturated fats — and certainly from butter.

What are we to make of this? How healthy or unhealthy is butter, then? If we consider the entire body of knowledge about butter, we can say, on the one hand, that butter, with its saturated fats, is undoubtedly less healthy than beneficial unsaturated fatty acids. But on the other hand, there appears to be no reason to ban butter from the kitchen completely. It might sound a little like fence-sitting, but that's precisely the point when it comes to butter. It could be described as kind of 'neutral'. And what does that mean? Well, it comes down to the question of what we would eat *instead*. Replacing butter with a good-quality rapeseed or olive oil would be a good call. If we eat white bread instead of butter … not so much. The authors of a recent large-scale analysis — taking in data from more than 630,000 people in different countries — summarise this important point as follows:

> Our results suggest relatively small or neutral overall associations of butter with mortality, CVD [cardiovascular disease], and diabetes. These findings should be considered against clear harmful effects of refined grains, starches [author's note: such as white bread, white pasta, white rice, and potatoes], and sugars on CVD and diabetes … In sum, these results suggest that health effects of butter should be considered against the alternative choice. For instance, butter may be a more healthful choice than the white bread or potato on which it is commonly spread.[6]

This might seem a little harsh on potatoes, but, on the whole, I think this conclusion is reasonable. It's interesting that the authors make the comparison between butter and white bread and potatoes, because it gives us a feeling for how much worse butter's reputation is than it deserves. For our day-to-day lives, it's probably more relevant to ask how butter measures up against other fat sources, such as olive oil, sunflower oil, and rapeseed oil, because they can often be used as a replacement for butter. And the answer to that is clear: butter is less healthy than those unsaturated oils.

I treat myself to butter maybe once or twice a week on average. I sometimes use it for frying and almost get drunk on its delicious aroma, and of course I use it on the (rare) occasions when I bake a cake. Sometimes I make myself a bulletproof coffee, just for fun. As described earlier, that's a cup of coffee with a tablespoonful of butter and two tablespoons of MCT oil mixed in. Whatever I'm making with it, I always use butter from the milk of free-range cows that have been allowed to graze on fresh grass. Like the milk it's made out of, this grass-fed butter contains slightly more omega-3 fatty acids and other beneficial substances. I can't actually prove it, but I imagine *this* butter really is healthy!

Cheese: source of vitamin K and cell-rejuvenating spermidine

Cheese is a similar case to butter, but more interesting and more complex. Cheese is richer in nutrients than butter. It's true that cheese also contains a high proportion of saturated fats, as well as protein. Compared to butter and other sources of saturated fats, eating cheese has a slightly more positive impact on blood-fat levels.[7] The reason for this is not well understood, but it may have something to do with the large amounts of calcium contained in cheese. In the gut, calcium binds to the molecules of the fat we've eaten, which stops so much of the fat from being absorbed by the gut. That means we simply excrete some of the fat we've consumed, thanks to calcium (this is not just theoretical speculation, but based on diet experiments that included, yes, examining the test subjects' stool).[8]

In addition, cheese is a source of some fascinating nutrients, such as vitamin K.[9] Vitamin K is best known for the role it plays in blood coagulation (the K comes from the German word *Koagulation*), but scientists have repeatedly been surprised to discover the many other important functions of vitamin K within our body.

For example, vitamin K protects our arteries from literally becoming calcified. It activates protein molecules that attach themselves to calcium, preventing it from accumulating in the walls of our blood vessels. The protein molecules thus actuated by vitamin K can also actively 'draw out' calcium from the vessel walls, decalcifying them. The calcium is then available for transportation to the places it's needed, including the bones, muscles, teeth, and brain. Junk food usually contains little to no vitamin K, and a deficiency of this vitamin can lead to calcium — an otherwise useful nutrient — accumulating in the artery walls, thus drastically increasing the risk of a heart attack.[10]

This interdependency between vitamin K and calcium is a good example of how taking individual dietary supplements can often backfire. Our bodies aren't keen on individually selected substances; they like nutrient *cocktails*, or nutrient *packages*. Or to put it another way, our bodies prefer real food to pills.

People who take in plenty of calcium *through their diet* (for example, by eating cheese, which also provides the necessary vitamin K) have a *reduced* risk of their coronary arteries becoming calcified. Any junk food fans who think they can simply redress their dietary balance by popping a few calcium pills are making a mistake. Due to the lack of vitamin K, the calcium will just gather in their blood vessels, calcifying them, including the coronary arteries, with the predictable negative — even fatal — consequences. A large-scale German study showed that taking calcium tablets increases the risk of a heart attack by no less than 86 per cent.[11] (There will be more on dietary supplements — which you need to take and which you don't — in chapter 11.)

So we know that it's important that we get vitamin K from our food. It may also help to prevent cancer; cellular studies have at least shown that it inhibits the growth of various cancer cells. Also, consuming food that's rich in vitamin K is associated with a reduced overall mortality risk, not least because of the reduced risk of mortality from cancer.[12]

There's a theory that the body switches to a sort of emergency mode when it's deficient in dietary vitamins and minerals, when it mainly uses those substances to secure short-term survival. For vitamin K, that means the body reserves it for blood coagulation. If an (internal) injury fails to heal and continues to bleed, the result is death, so this use of vitamin K has absolute priority. The fact that it leaves no vitamin K available to prevent a gradually advancing case of calcification of the arteries is of secondary importance, from an evolutionary point of view. In harsh terms, nature doesn't care if our coronary arteries will be calcified by the time we're 50. It cares about making sure we don't bleed to death from a simple wound before we reach that age!

When the body is deficient in a vitamin such as K, the resulting physical problems aren't acute — we only begin to notice them gradually. They manifest themselves as we get older in the form of osteoporosis, cardiovascular disease, or cancer. In other words, the vitamin and mineral deficiencies that result from a junk-food diet do not kill us immediately, but they accelerate the ageing process.[13] The bill doesn't come till years later. Eating foods that contain vitamin K could be seen as an investment for your old age.

An unrivalled source of vitamin K, by the way, is the Japanese dish of fermented soybeans called *nattō*, also known as 'vegetarian stinky cheese' (caution: *nattō* more than deserves that name and is certainly not to everyone's taste, although it is a real heavyweight when it comes to health benefits).[14] Most of us would prefer to eat actual cheese. It may not be of quite the same nutritional calibre as *nattō* — but at least it doesn't make other people angry when you eat it. However, eating cheese does make many of the people eating it feel guilty, unfortunately. That's a real shame because, all things considered, eating cheese is to be thoroughly recommended.

As well as calcium and vitamin K, cheese contains at least one other very remarkable and healthy substance: spermidine. Its rather unfortunate-sounding name comes from the fact that it was originally isolated from semen. In fact, spermidine is found in practically all our body's cells. The concentration of spermidine in our cells decreases as we get older, although, interestingly enough, this is not the case for people who live to an unusually great age (there's a remarkably large amount of spermidine in the blood of centenarians). Is spermidine a fountain of youth?[15] There might be some truth in that. In a similar way to rapamycin, spermidine has been found to extend the life span of several organisms and animals. And like rapamycin, spermidine

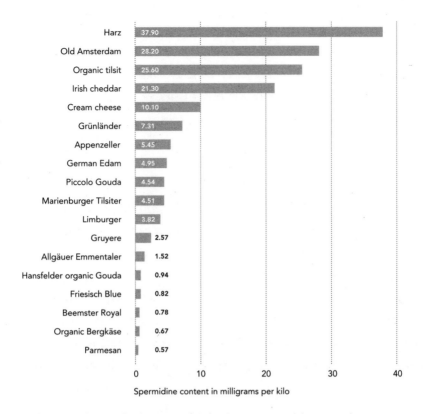

Fig. 9.1 Researchers at the University of Kiel in Germany tested the spermidine content of 50 different cheeses commonly available in my homeland. This is an excerpt from their results. As you can see, the concentration of spermidine varies greatly, which has several contributing factors, including the bacteria and enzymes in the raw milk, heat treatment, and the different maturing times of the various types of cheese.[16]

also activates the cell's self-cleaning program (autophagy) and thus rejuvenates the body from the inside out.[17]

The great thing about spermidine is that it's contained in many of the foods that should already be part of a healthy diet, and eating those foods generally enables the spermidine they contain to be taken up well by the body, so that its positive effects can unfold.[188] For example, people who eat a lot of spermidine-rich foods have a 40 per cent lower risk of suffering fatal heart failure (compared to those who eat no spermidine).[199] The biggest spermidine source is wheatgerm (that part of the corn seed that develops into a new plant, incidentally also an excellent source of plant protein and very tasty). Other good sources are soybeans, mushrooms, peas, broccoli, cauliflower, apples, pears, lettuce, wholemeal products, and, of course, cheese, although the content can vary greatly from type to type.[20] As a rough rule, more mature types of cheese contain more spermidine than younger kinds. However, that rule doesn't always apply, as you can see in fig. 9.1. Top of the spermidine charts, Harz cheese, for example, matures in just a few days. And parmesan, which matures over a period of months, contains hardly any spermidine at all.[211]

Conclusion: cheese is fine, cheese is our friend.[22] I don't want to sing the praises of butter and cheese too highly, but they are certainly better than their reputations. Our readiness to condemn butter and cheese so quickly highlights how counterproductive it can be to demonise traditional foods without the basis of sound scientific research. The simple presence of saturated fats is not a sound basis, and such demonisation led the food industry to give us modern margarines, whose trans-fat content turned out to be really toxic. Even now, many of us deny ourselves our beloved camembert, only to replace it in our diets with rapid carbs or some sort of fat-free (= sugary) industrial snack, to no ultimate advantage. The opposite, in fact.[23] No more! I wish all cheese lovers bon appétit!

CHAPTER 10

Fats III:

oily fish and omega-3 fatty

acids as slimming food.

Or: food as information

Fish and fishoids

Two of the most popular kinds of fish in Germany are salmon and Alaska pollock.[1] Apart from their popularity, the two fish have very little in common. Most of us are vaguely familiar with what a salmon looks like. But Alaska pollock? Germans have no idea because we only ever see it in the distorted form of an industrial product — as a deep-frozen bake with a palm-oil, glucose-syrup, and sugar topping. Or as deep-fried, breadcrumbed fish fingers.[2] More fishoid than fish. But before it gets the commercial treatment, Alaska pollock is a species of the cod family.

The important thing for us is what's in the fish. Of course, fish is also a package of many nutrients and can't be reduced to one single substance. Still, in this case, there can be little doubt that the famous omega-3 fats that fish contains gives it special value as a food.

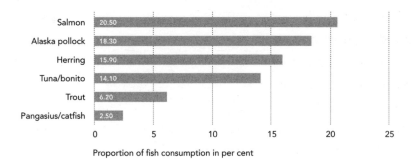

Fig. 10.1 Alaska pollock is officially one of the most popular fish among Germans — however the way industry 'prepares' it for us leads me to use the word 'fishoid' for this finished product.[3]

The ultimate source of omega-3 is the plant world. Omega-3 fatty acids can be found, for example, in grasses and seeds. Fish get them from algae. Those fatty acids are often somewhat lacking in our modern diets since we eat so many different animals and animal products these days, and most of the animals themselves never get to graze on open pastureland, being kept indoors and fattened on concentrated feed with little to no omega-3 content. You are not only what you eat. You are also what the things you eat ate. Animals who have no sources of omega-3 in their diet cannot provide omega-3 in our diet.

There are different kinds of omega-3 fatty acids, and fish is the best source of some particularly healthy types of omega-3 — but not all fish. Only oily fish provides us with omega-3 in any meaningful amount. That means salmon, herring, tuna, trout, sardines, mackerel. Scampi and shellfish contain smaller amounts of omega-3.

Salmon is one of the richest sources of omega-3 — especially farmed salmon, incidentally. There's a myth that farmed fish contain less omega-3 than those caught in the wild. In reality, it's actually the other way around. A farmed salmon or trout contains significantly more of the fatty acid, although this also depends on what the fish have been fed. Because farmed fish is generally so much fattier than wild fish (aquaculture fish receive large amounts of food without having to expend energy to get it), the *proportion* of omega-3 in the overall fat content is lower in farmed fish than wild ones, but in absolute numbers, farmed fish have more omega-3.[4]

Personally, I prefer wild salmon, but I also often eat farmed salmon because fresh wild salmon is so difficult to get hold of. The fresh trout in your local supermarket also almost always comes from some farm or other. Unfortunately, the conditions under which the fish are kept in such farms are abysmal (overcrowding, prophylactic use of antibiotics, low-quality feed, and more), which ultimately has to do with the fact that we expect even the complex biological creatures we eat to cost nothing. They are literally worth nothing to us! In my opinion, fish *should* be a little bit more expensive (as should all kinds of meat and animal products). I've already said this earlier in the book, but people used to speak of the 'Sunday roast'. Nowadays, we're disappointed if we don't find meat on our plates every day. Maybe it wouldn't be such a bad thing if meat and fish (once again) became something that people can only afford to eat every now and then. By the way, that would also do our health a lot of good, because, healthy as oily fish is, we don't need to eat all that much of it to feel its positive effects.

Compared to other popular types of fish, such as herring, tuna, and trout, which are also rich sources of omega-3, Alaska pollock contains relatively little omega-3. Pangasius, which has gained vastly in popularity in recent years, has almost no omega-3 fats at all (see fig. 10.2). But it is full of mercury and other toxins.[5] Pangasius comes from fish farms in Asia, mostly in Vietnam, where 'animal welfare' is a little-known concept. The fish are kept in extremely cramped conditions in which they are unable to swim around (imagine 40 good-sized fish in your bathtub).[6] I do not recommend eating this fish.

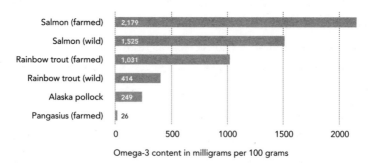

Fig. 10.2 Despite common claims to the contrary, farmed salmon doesn't contain less omega-3 than wild salmon, but considerably more, and the same is true of rainbow trout. Farmed fish is generally much fattier than wild fish. Some fish, such as pangasius, which is practically always farmed, contains almost no omega-3 at all.[7]

How the fish we eat becomes part of our cell membranes

Why are omega-3 fatty acids so good for us? The short answer is because they are far more than just a simple energy source. For the more complex answer, I need to go back to biological basics a little.

We need food because we need energy. But food provides us with more than just power. Especially when we talk about fats, we usually think immediately of the calories they contain. Our fixation with calories means we lose sight of the fact that fats are more than just well-stuffed packets of calories. As you will see, the *informational character* of omega-3 fats, for example, means they can even be helpful in losing weight. As I've already explained, some of the fatty acids we eat never get 'burned' for their energy. Rather, they become a part of us. The fatty acids are integrated into the membranes of our cells, which makes them either stiffer or more supple, depending on the fatty acid in question. This can change the whole way the cell works.

A particularly eye-catching example of this process is, in fact, in our eyes. The form of omega-3 found in the terrestrial plant kingdom — as mentioned before, present in grasses, linseed, and chia, but also in walnuts and rapeseed — is known as alpha-linolenic acid. This omega-3 fatty acid is made up of a chain of 18 carbon atoms with three kinks in it. So it's a polyunsaturated fat.

When we eat alpha-linolenic acid, this fatty acid is modified within our bodies, primarily in the liver. There, alpha-linolenic acid can first of all be lengthened. This means that some more carbon atoms are attached to the end of the chain. Secondly, the carbon chain can be given *even more kinks*, making it *even more unsaturated*. Women's and children's bodies are better at doing this than men's, which tells us that the resulting, extended, highly unsaturated omega-3 fats are particularly important for the development of (unborn) babies.

But we can also eat these long, highly unsaturated omega-3 fats directly, and the direct source is oily cold-water fish such as salmon and trout. The highly 'kinky' fatty acids remain liquid even at low temperatures — even below zero. In salmon, the highly unsaturated omega-3 fats have a kind of anti-freeze effect, keeping the fish's body supple even in extremely cold conditions. This presumably gave salmon an advantage in the struggle for survival, preventing their bodies from solidifying like a block of butter in freezing cold sea or river waters, and giving them more energy to move in the cold (salmon live in both marine environments and rivers, where they return to spawn — i.e. to mate and lay their eggs).

Put simply, it is mostly those long, strong unsaturated omega-3 fatty acids derived from salmon and other oily fish that become embedded in our cell membranes. The most important members of this group of omega-3 fatty acids are called EPA (eicosapentaenoic acid), DPA (docosapentaenoic acid), and DHA (docosahexaenoic acid).

The eye and the brain in particular contain large amounts of DHA. DHA is unusual, even among the long omega-3 fatty acids, for having so many kinks (six in all). They give the fatty acid a circular shape, like a snake biting itself on the tail. A DHA molecule looks like a little loop. This molecule's very 'airy' structure has a unique influence on our cell membranes.

In our eyes, our retinas contain light-sensitive cells, some of which (the 'rods') enable us to see in the dark, at twilight and at night. This is how they do it: the (fatty) cell membranes of the rods contain molecules of a protein called rhodopsin. As soon as light hits one of these rhodopsin molecules, it changes shape, prompting the rod cell to send a message to our brain ('Light!'). This is how we see.

The rhodopsin molecules, for their part, are surrounded by the fatty acids that are incorporated in the cell membrane. That means the cell membrane is made up of fatty acids peppered with rhodopsin molecules. It's as if the rhodopsin molecules were floating like buoys in a thin layer of fat (= cell membrane). Depending on what we eat, this layer is made up of different fatty acids, and that has an effect on the function of the rhodopsin molecule.

It turns out that, unlike other fatty acids, DHA supports the light-stimulated change in shape of rhodopsin molecules so as to drastically improve the transfer of their signals. Imagine doing gymnastic exercises first in a diving suit and then in a loose-fitting T-shirt and tracksuit pants. Which is more fun? It's a similar situation to that of rhodopsin in the membranes of the cells in our retina. There, rhodopsin can do its exercises far better in the 'tracksuit' of the 'airy' omega-3 fatty acids that surround it. Omega-3 fatty acids act like a kind of activewear for the retina. This is why omega-3 fatty acids are so important for our eyes, and the same is true of the development of a baby's eyes while it's in the womb. In other words, the fish we eat, or that our mothers ate, is not just burned up for energy; rather, some of it ends up in our eyes, helping us to see clearly.[8]

Nearby, many omega-3 fatty acids are incorporated into our brains, which are extremely fatty organs in general. When we're 30-week-old embryos, our brains weigh only about as much as a mandarin orange (100 grams). Not much more than a year later, by the time we're 18 months old, our rapidly growing brains have gained an entire kilo in weight, now weighing in at 1,100 grams. In the same period, the amount of DHA they contain has increased by *35 times*![9]

Just as they are in the eye, DHA and other omega-3 fatty acids are very helpful to the functioning of the brain. Transferring signals is the be-all-and-end-all of the brain's modus operandi. It's the uninterrupted exchange of information between the nerve cells (neurons) that drives our entire inner world, our thoughts, our fantasies, our emotions, all the way to the thing we call 'me'. When the transfer of signals in the brain is well facilitated by the omega-3 fatty acids built into its neurons, we not only see more clearly, but also think more clearly.

A German study with the participation of Berlin's Charité hospital found that even a simple course of fish-oil capsules (four capsules a day, containing a total of around 1.3 grams of EPA and 0.9 grams of DHA) can powerfully rejuvenate the brain structure of adults between the ages of 50 and 75. The study ran for six months. In that time, the brain volume of the control group shrank noticeably, by just over 0.5 per cent. This 'usual' brain shrinkage was halted in the people who were randomly assigned to the group that received omega-3 capsules. There was even an *improvement* in the structure of some regions of their brains. So it's no wonder that the omega-3 test subjects outperformed those from the other group in all kinds of cognitive tasks. The more omega-3 fatty acids had accumulated in the body of the test subjects, the more their word fluency, for example, improved. (How many words can you think of that begin with the letter S? The more words you can come up with in the space of one minute, the better your word fluency is.)[10]

But we don't just think with our brains, we also *feel* with them. That's why omega-3 fatty acids can affect our mood and even potentially considerably improve it. Patients with depression are often found to be suffering from a lack of omega-3.[11] What's more, the level of severely depressed patients' DHA deficit is an accurate indicator of their future suicide risk![12] By the same token, a daily dose of fish-oil capsules (4 grams of fish oil in total, of which 1.6 grams were EPA and 0.8 grams DHA) has been found to have a positive influence not only on the brain structure of depressed patients, but also on their depression in general.[13] One mechanism that may be key to all this is the one by which omega-3 fatty acids are able to promote the generation of new brain cells in a structure called the 'hippocampus'. Like most of us, I was taught at school that the brain doesn't grow any new neurons after birth. That turns out to be wrong. It is possible in certain regions of the brain, including the hippocampus. One of the important functions with which this constant supply of new neurons probably helps us is learning (ironically enough, perhaps even in learning the myth that no new neurons can grow in our brains after birth).

The hippocampus is crucial in the creation of new memories. Oddly, the hippocampi of people with depression are also sometimes smaller than normal. Depressed people sometimes consult their doctor initially about problems with their memory. Magnetic-resonance imaging can then reveal that the hippocampus of such patients has *shrunk*.[14] Omega-3 fatty acids can help their shrunken hippocampus to grow back by stimulating this brain structure to grow new nerve cells. It's as if the brain structure had been 'cured'. It might sound like science fiction, but the effect has been registered not only in many animal tests, but also in tests on us humans.[15]

One speculation in connection with this is that unborn babies need so much DHA in the last phase of pregnancy that they draw omega-3 fatty acids from their mother's bodies if they're not getting enough through the mother's diet. The unpleasant side effect of this is an omega-3 deficiency in the mother, which may contribute to the mood changes often experienced by new mothers, ranging from a case of the 'baby blues' all the way to clinical postpartum depression (see also the striking correlation shown in fig. 10.3).[16] I'm not suggesting that all perinatal depression is due to a lack of omega-3; there are many contributing factors (hormonal, psychological). But it's possible that some cases can be perceptibly improved by a few extra fish- and plant-based omega-3 fatty acids.

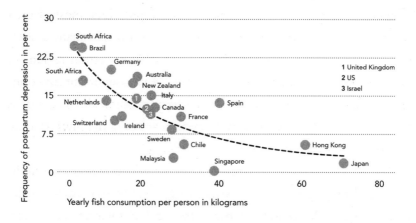

Fig. 10.3 The lower the fish consumption in a particular country, the higher the number of women who report suffering from postpartum depression. Of course, this observation does not prove a causal link. One hypothesis is that unborn babies need so much omega-3 fatty acid to build their brains that they draw it from the bodies of their mother (unless the mother eats a lot of fish). The resulting omega-3 deficiency in the mother then increases her risk of depression.[17]

How fish instructs our cells

Omega-3 fatty acids don't change our bodies and our minds only by being incorporated into our cell membranes and optimising the function of the molecules embedded in them. Fatty acids can also bind to certain receptor molecules in the cells of our body and pass on signals to them in that way. The omega-3 fatty acids we eat 'speak' a molecular language to our cells. They provide *information* — with beneficial effects for our health.

You may remember from chapter 4 that fats are conveyed via the bloodstream to our organs with the aid of transport capsules. Those capsules stop the fat from gathering in clumps in our blood. There are also so-called free fatty acids floating around in our blood that are bound to just one protein molecule — again, to stop them from accumulating in blobs. These free fatty acids are among the ones that latch onto our cell membranes and can then affect the inner lives of our cells.

Let's take intra-abdominal fat as an example. It secretes inflammatory substances just like a gland: some of those inflammatory substances are secreted directly by the fat cells themselves, and some are secreted by the phagocytes (scavenger cells) of the immune system that infiltrate the intra-abdominal fat. However, both the fat cells and the phagocytes of the body's defence system have receptors on their outer surface that function as omega-3 sensors. When a free DHA fatty acid binds to such an omega-3 receptor, it triggers a cascade chemical reaction inside the fat cell or phagocyte that results in many genes being activated or deactivated. The positive consequence of this intervention in the cell's genetic activity is that it halts the — excessive — formation of inflammatory substances. The inflammation is curbed. It's almost as if the omega-3 fatty acids were acting like a kind of healing ointment for an internal injury. This means the trout that looks so lifeless before being devoured for dinner becomes part of the cell membranes in our eyes and brains, and also gives instructions to our genes that result in a reduction in the very aggressive inflammatory processes in our body.[18]

I can't emphasise enough how important that reduction of inflammatory processes is. The typical diseases of old age, from rheumatism and atherosclerosis to dementia and cancer, are all associated with chronic inflammatory processes and are expedited by the substances produced as a response. Presumably, the ageing process as such is also accelerated in the same way. By the same token, gently inhibiting such inflammation could slow down the ageing process; there are early, but spectacular indications that this may be the case.

In a large-scale study published in the journal *Nature*, researchers at the Albert Einstein College of Medicine in New York showed that the ageing process in mice could be accelerated or decelerated almost at the flick of a switch, simply by activating or inhibiting one of the central inflammatory switching substances, called NF-κB, in the brains of the mice. When microbiologists talk about inflammation, they are usually talking about NF-κB. It's like a general in the army that defends our bodies (or perhaps an admiral would be a better comparison, since our bodies are so water-based). When NF-κB is activated, this commander-in-chief of the immune system launches a molecular military campaign. NF-κB controls hundreds of genes that massively mobilise our body's defences. This is very useful in acute situations, such as when we catch a cold or injure ourselves. Those are the situations in which our immune system is supposed to defend us and keep everything in good order. It's only when a mission becomes drawn out, takes on a life of its own, and becomes a permanent state of affairs — as is so often the case when we are old or overweight — that the collateral damage gets out of hand. The permanent military mission carried out by our immune system begins to damage our own bodily tissue, destroying it, and causing the body to age more quickly.

Activating NF-κB in the region of a mouse's brain called the 'hypothalamus' — a small but very influential structure of the brain that regulates the processes of growth, reproduction, and metabolism, and also acts as the brain's 'fullness centre' — has the result of inhibiting certain hormones. This hormone-low in turn accelerates the ageing process throughout the mouse's *entire* body, and muscles atrophy, osteoporosis occurs early, skin tension is lost, physical fitness is reduced, and memory performance diminishes. The mouse ages faster and dies earlier. Contrariwise, all those effects can be prevented, and the mouse's life prolonged, by doing nothing more than inhibiting the inflammation commander — NF-κB — in the hypothalamus.

This is a truly astonishing discovery. It suggests that ageing and physical decline are not the inevitable consequences of a process of wear and tear, as we imagine, analogous to the clapped-out old car we drive. Rather, ageing can be seen as a program that our brain runs, similar to puberty. If our brain (or more specifically, our hypothalamus) is 'inflamed', our whole body ages more quickly.[19]

There's some good news in all this. If ageing is a process that's controlled by the brain, it may be possible to reprogram the brain to halt that ageing process. And, indeed, it seems to be possible, at least to some extent.

At least, it's possible to do something about the inflamed state of the brain. Interestingly enough, the cell membranes of the hypothalamus are also equipped with omega-3 sensors. This is the pathway via which omega-3 fatty acids can reduce the inflammation of the hypothalamus,[20] which should have a positive influence on the ageing process throughout the whole body. This is all still very speculative at the moment, but if there turns out to be a kernel of truth to it, it would mean that eating fish regularly could offer protection against ageing too quickly.

Furthermore, there are indications that the anti-inflammatory effects of omega-3 fatty acids may be helpful in losing weight. Obesity *also* often leads to an inflammation of the hypothalamus, which can massively disturb its function. Thus, obesity is not only associated with inflamed intra-abdominal fat, but also with an inflamed brain, or at least hypothalamus. Since it's the hypothalamus that's responsible for our feeling of fullness after eating, one of the consequences of obesity can be losing some of our ability to feel full. We feel permanently hungry, not despite, but *because of* our all-too-ample fat reserves. How does that work? Simple: just as an inflamed nose can barely smell anything, an inflamed hypothalamus can barely 'smell' the signals coming from our body telling it we've eaten enough. It may be unpleasant to have an inflamed nose due to a cold, but at least we notice that it's inflamed. If our hypothalamus is inflamed, we don't notice a thing, at least not directly, as the brain itself is numb.

This is how obesity can cause ever more obesity, because the 'blocked-up' hypothalamus is no longer able to perceive the ample energy supply in the body. This vicious circle can be broken by omega-3 fatty acids and their anti-inflammatory effect on the hypothalamus. It can then once again register all the calories we eat, and we stop constantly feeling hungry.[21] Scientific findings show that both eating fish and taking omega-3 capsules can help with losing weight.[22]

In summary, oily fish and omega-3 fatty acids are to be recommended — in moderate amounts. Eating fish lowers the risk of many diseases associated with old age, from cancer and cardiovascular disease to cognitive decline.[23] Fish especially, but possibly also omega-3 capsules (this is currently being studied extensively[24]), may reduce the overall risk of mortality.[25]

Oily fish and omega-3 capsules, perhaps unsurprisingly, have a beneficial effect on inflammatory diseases such as the painfully inflamed joints often suffered by people from the age of about 40 or 50 (rheumatism, or, more precisely, rheumatoid arthritis).[26] Fish and fish-oil capsules are no cure-all. But they can provide a valuable contribution to a healthy lifestyle.

My recommendation is, as I already mentioned, to eat one to two portions of oily fish per week. Those who don't like oily fish might try omega-3 capsules as an alternative, especially if they are overweight or have been diagnosed with inflammatory conditions by their doctor.[27] The usual dose for fish-oil capsules is two to a maximum of three per day. A capsule usually contains one gram of fish oil. A little over half of that is a mixture of different omega-3 fatty acids, usually mostly EPA and DHA. My tip: there are molecularly purified fish-oil capsules that are largely free of any contamination with mercury or other harmful substances. One alternative to fish oil, which presumably has the same effects, is krill oil.[28] Another alternative is algae oil, which is suitable for vegans. The capsules should be kept in the refrigerator, otherwise the oil they contain can become rancid.

Fats: summary and *Compass* recommendations

Fat — the mere word proclaims our fate when we eat from this third and final major food group. Fat makes you fat, as the cliché goes. Indeed, at nine calories per gram, fat does have more energy than carbohydrates or protein, which provide us with just four calories per gram (pure alcohol is somewhere in between, at seven calories per gram). This is compounded by the fact that we intuitively imagine our blood vessels to be like 'drainpipes', which are notorious for getting blocked by fat. Against this background, it's no surprise that the demonisation of fat attracts so many followers. This fatphobia has led to us eating more and more rapid carbs and industrially produced sugary foods, some of which have turned out to be far more harmful than fat.

Today we know that most fats are harmless and some are actually good for us. Particularly beneficial are omega-3 fatty acids, found principally in linseeds, chia seeds, walnuts, rapeseed oil, and oily fish (e.g. salmon, herring, trout). The amount of energy it packs is not the only thing about what we eat that is important for our health, or even our figure. The physiological effect of the food is far more important. Some of the fatty acids we eat are incorporated into the structures of our body — our cell membranes — and also function as signalling substances, which can, for example, reduce inflammatory processes.

The 'medical signalling factor' of many fatty acids turns out to be a blessing for obese and old people. The 'drainpipe model' is not really an accurate way of describing the biology of our blood vessels. Rather, atherosclerosis is, importantly, an *inflammatory* disease. LDL particles accumulate in the vessel walls where they begin to 'rust', causing an inflammation. This explains why omega-3 fats not only don't clog our arteries, but, on the contrary, have an anti-inflammatory effect, thus *lowering* the risk of cardiovascular disease. Omega-3 fats also lower the risk of many other diseases associated with ageing, including, for example, rheumatism, which is also an inflammatory disease. Since obesity is also associated with an increased inflammatory response, omega-3 fatty acids can have a beneficial effect here, too.

And equally importantly, healthy fats are also our friend when it comes to insulin resistance. Since insulin resistance increases with age, it may generally be a good idea to eat a little less carbohydrate and a little more fat later in life. Personally, I definitely eat more fat than I used to, mainly in the form of linseeds, nuts, olive oil, rapeseed oil, avocados, dark chocolate, and fish. I also eat a little more cheese than I used to.

As a rule of thumb, we can state that unsaturated fats are healthier than saturated fats. Yet even saturated fatty acids are fine, by and large,

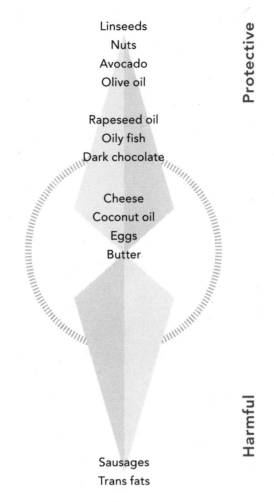

Linseeds
Nuts
Avocado
Olive oil

Rapeseed oil
Oily fish
Dark chocolate

Cheese
Coconut oil
Eggs
Butter

Protective

Harmful

Sausages
Trans fats

Fats compass needle

Contrary to popular belief, most fats are basically healthy, and some fat-rich foods are even highly recommended. The only type that should be avoided at all costs are the trans fats — they are toxic.

especially in cheese. Butter can be considered 'neutral', as can the currently hyped coconut oil. The reason for this hype is that coconut is wrongly considered to be an MCT oil — consisting of healthy medium-length saturated fatty acids. MCTs, however, make up only 15 per cent of coconut oil. Please don't misunderstand me here: coconut oil is fine, it's just not a 'super food'.[29]

The fatty foods you should really avoid at all costs are sausages, as well as — because of the trans fats they contain, among other things — donuts, potato chips, fries, and other deep-fried foods, readymade pizzas, and any baked goods that weren't made by your granny (i.e. industrially produced ones).

CHAPTER 11

No vitamin pills! Except ...

For everyone: vitamin D

We don't need vitamin pills — well, not many. If you like to drink a fresh smoothie or multivitamin juice, great, but you should limit yourself to one a day. Don't think of juice as a replacement for real fruit in your diet.

Most vitamin pills are 'just' a waste of money, but some are actually downright harmful. Taking vitamin A and beta-carotene (a precursor to vitamin A), for example, in concentrated pill form can even increase the risk of mortality.[1]

According to current scientific knowledge, there's only one vitamin that can lower the risk of mortality when taken in pill form — vitamin D — and it's a special case in many respects. As already described, our bodies like to get their nutrients in packets, not individually. By the way, anyone who even vaguely follows the recommendations I make in *The Diet Compass* can rest assured that they are getting all the vitamins and minerals they need, and in sufficient amounts. But vitamin D is an exception, for two reasons.[2]

Firstly, vitamin D is found in very few foods. Essentially, those are oily fish (such as salmon, mackerel, and herring), cod-liver oil, and mushrooms once they have been exposed to sunlight or have been dried in the sun.[3] Our body simply produces most of the vitamin D it needs by itself, which is why D is, strictly speaking, not actually a vitamin at all. A vitamin is defined as a substance that the body requires in small quantities but can't produce itself. However, as you probably already know, our bodies can only manufacture sufficient vitamin D if our skin is exposed to enough sunlight (or, more accurately, to ultraviolet-B radiation, UVB).

One reason why some members of our species evolved lighter skin may be that it helps the body make more vitamin D. Darker skin pigments act as a barrier to UVB radiation, making dark skin naturally protective against sunburn. That was practical in the savannahs of Africa, which were once home to the entire population of *Homo sapiens*, and it's still practical there today. Yet the further from the equator you go, for example towards the north, the more difficult it becomes for darker skin to synthesise vitamin D. In Germany, the light of the sun is so weak in wintertime that even light skin is unable to make any vitamin D — no matter how much time white-skinned Germans spend outdoors.

This has led to a serious vitamin-D deficit in Germany, as revealed by several scientific investigations. Some experts consider 50 nanomoles or more of vitamin D per litre of blood to be ideal, but a survey of all the relevant research findings strongly indicates that the optimum level is more likely to be 75 nanomoles per litre and above.[4] Whichever level you take, the fact is that German vitamin-D levels are too low. As you can see in fig. 11.1, my kinfolk don't reach the level set as *low, even in summer!* It's not an overstatement to say that, from the point of view of optimal provision, there's almost a nationwide epidemic of vitamin-D deficiency in Germany[5] — and this is true of one billion people worldwide, in all ethnicities and age groups.

In short, we need vitamin D, and when I say 'we', I mean every single one of us, although some people need more than others. There are two kinds of vitamin-D supplements — D_2 and D_3. D_3 (cholecalciferol) has been found to be the more efficient of the two. It's also the form of vitamin D synthesised by our skin and contained in fish.

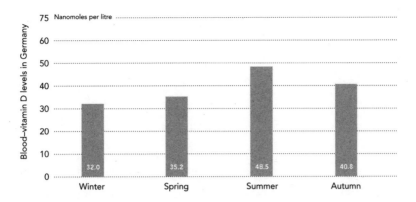

Fig. 11.1 Vitamin-D levels in Germany are well below the optimum (75 nanomoles per litre or more), even in summer.[6]

We've known for decades that the main function of vitamin D is to enable the body to take in calcium. That's why vitamin D is important for strong bones and prevents bone deformities in children, a condition known as rickets. Recent research has shown, however, that almost every organ in our body is equipped with vitamin-D receptors. This shows that vitamin D plays many roles throughout our body, and those roles are far from being completely understood. For example, it was only recently that scientists discovered that vitamin D can protect against the common cold.[7] Which might explain why we are so much more liable to catch a cold in winter, when our vitamin-D levels are at their absolute lowest. Incidentally, a painstaking analysis of 56 scientifically sound studies showed that taking vitamin-D_3 supplements can protect against an early death.[8] The most recent analyses show that vitamin-D_3 supplements reduce the risk of mortality by no less than 11 per cent![9]

Taking D_3 is good, then. But what's the best amount to take to achieve the blood levels recommended above? Well, that depends on your personal situation. As a guideline, we can say: 1,000 to 2,000 international units (IU) per day are sufficient for most adults (1,000 international units equals 25 micrograms). The maximum daily dose judged to be safe for adults is 4,000 units. This daily amount should not be exceeded.[10]

If you spend a really great deal of time in the summer sun, then you probably don't need any supplements at all, especially in the last months of summer. Your body will have accumulated enough vitamin D naturally. If this is the situation you're in, you should start taking 1,000 units mid-autumn, increasing the dose to 2,000 in winter, and gradually reduce it again as spring approaches and the sun comes knocking on the door once more. This is the strategy I follow.

A word to sun worshippers: not only should you never let your skin get burnt, but you should get out of the sun even before your skin starts to redden. The best thing is to 'spread' the sunlight equally over your whole body. That means it's better to spend a quick 20 minutes in the sun — naked, as far as possible — in the afternoon, than to fry your face alone for an hour in the aggressive sunlight. In summer, I use sunscreen with a sun protection factor (SPF) of at least 30 on my face (don't forget your ears!) and the back of my neck every day, which theoretically means I could stay in the sun 30 times longer. I usually also wear a hat.

Some people need more vitamin D than others:

- Many older people spend very little time out of doors, so vitamin-D deficiency is particularly high among the elderly. This is compounded by the fact that old people's skin synthesises less vitamin D. In this case, a dose of 2,000 units per day is recommendable all year round.
- As vitamin D is a fat-soluble vitamin, one of the places the body can store it is in fat cells. The more fatty tissue there is, the more of the vitamin is 'swallowed up' in this way. So people who are (severely) obese need more vitamin D.
- The darker your skin, the more vitamin D you need.
- Big-city dwellers tend to have less exposure to the sun than country folk. So city slickers who spend the time from 9 to 5 cooped up in the office need a little more vitamin D. (There's also some difference between latitudes here:[11] people

who live further from the equator tend to need to take more supplements than those who live closer.)

For vegetarians and vegans: vitamin B$_{12}$ at least

B vitamins are typically made by plants, with the exception of vitamin B$_{12}$. This special vitamin is produced by bacteria. It is almost completely absent from plants.[12] That means you should take vitamin-B$_{12}$ supplements if you're a vegetarian, and especially if you're a vegan. The recommended amount is 250 micrograms of cyanocobalamin daily.[13] This is true to an even greater extent for pregnant and nursing mothers if they are vegan, because a vitamin-B$_{12}$ deficiency can cause serious neurological damage to an unborn baby.[14] This isn't something to be taken lightly! A vitamin-B$_{12}$ deficiency can completely reverse all the advantages of a vegetarian or vegan diet. If you don't eat fish, I would also recommend taking omega-3 supplements (a vegan version made of algae oil is available).

New research suggests the different B vitamins — B$_1$, B$_2$, B$_3$, B$_5$, B$_6$, B$_7$ (biotin), B$_9$ (folic acid), and B$_{12}$ — work together to a great extent, like the different instruments in an orchestra. Taking large doses of *one* particular vitamin, for example folic-acid tablets, can actually make a vitamin-B$_{12}$ deficiency worse. One vitamin is 'drowned out' by the other. It's like when the orchestra's double basses are too loud and drowning out the other instruments, ruining an otherwise harmonious symphony.[15] It's still speculation at the moment, but this might be a reason to take a well-balanced 'vitamin-B complex', uniting the entire family of B vitamins in one tablet, rather than a vitamin-B$_{12}$ supplement alone. B vitamins are water-soluble, which means taking too many results only in expensive, vitamin-rich urine.

Now I am definitely entering the realm of the purely speculative, but taking a vitamin-B complex may turn out to be beneficial for us all. B vitamins (unlike vitamin D) don't lower the overall risk of mortality, but they do at least lower the risk of strokes by 12 per cent.[16]

B vitamins may also be able to halt the gradual degeneration of the brain that is unfortunately typical in old age. A study carried out by the University of Oxford entailed giving elderly test subjects with memory impairment high-dose B vitamins over a period of two years. While the brains of the control-group participants shrank continually, this loss of grey matter was almost completely halted in the test group that received the B vitamins.[17] However, the effect was not universal. A further analysis revealed that B vitamins only protected the brains of those who had high omega-3 levels in their blood.[18] Firstly, that's yet more evidence in favour of a diet rich in omega-3. Secondly, it shows once again that our bodies like complex packets of nutrients, and we do not know the ideal combinations. One thing is clear, however: the best way to get a balanced combination is to eat a balanced diet.

I take 1,000 to 2,000 units of vitamin D a day; a little more in winter, a little less, or even none at all, in summer. I now go out in the summer sun more than I used to. I usually eat fish at least once a week, as well as eating linseeds every day and walnuts regularly. If I'm not eating fish for an extended period, which occasionally happens, I take two omega-3 capsules a day, just to be sure. In addition, I often take a vitamin-B complex — as a lover of wine and my brain. I don't plan to change this regimen in future, unless new information becomes available with better advice (which I fully expect to happen). That's why it's a good idea to keep a keen eye on the results of the latest big studies (for example, Harvard University is currently carrying out a large-scale study on the effects of vitamin D_3 and omega-3, both individually and in combination — the results are eagerly awaited).[19] My conclusion: I appreciate the few little tablets I take, but in general I prefer really complex nutrient symphonies. Here's to real food.

CHAPTER 12

Timing your eating, and the most effective way to fast

Why it's important *what* you eat, *when*, and for *how long*

Take a look at these two cute little creatures:

Two mice from the same genetic strain. The same size, the same age, and — here comes the surprise — both fed the same diet. Not just the same food, but the same amount, too. How is that possible? Why is the mouse on the left fat and the one on the right thin?

Imagine these two mice were people. We know nothing about them. We can just see that one person is fat and the other is thin. What would be the first thing to go through your mind? It would be difficult not to suspect the person who is fat had eaten more than the one who is thin. But, I repeat: both mice received exactly the same food.

Incidentally, the two were fed a kind of fast food for mice. So it's not completely astounding to see that the one on the left is fat. The remarkably good figure of the mouse on the right requires explanation. Why is it so slim? Was it maybe given lactobacilli along with its food, or some kind of magic elixir that miraculously protected it from getting fat? No.

That leaves only one other conventional explanation. The mouse on the right must have followed a program of rigorous exercise, while the one on the left led the life of a couch potato. But that isn't the case, either.

The actual answer is as simple as it is surprising — and it gives us a clue to a potentially extremely useful tip for losing weight. The fat mouse on the left was allowed to eat its junk food anytime it wanted, right round the clock. The slim mouse on the right was only given access to its food at certain, very limited times at night, when mice are naturally active. More precisely, the mouse had access to its food for eight hours a night. In the other 16 hours of the day, it was forced to fast.

Now, mice are clever, and, in situations like that, they quickly learn to stuff their bellies quickly during the 'short' periods when food is available. The result is that they ultimately eat the same amount of food as those who have permanent access to it. Nonetheless, they stay slim. And that's not all: they remain astonishingly fit into old age, which is all the more remarkable considering the unhealthy diet they were placed on. By contrast, the mice who were allowed to snack whenever they wanted not only grew increasingly fat, but also developed the typical conditions associated with old age in our well-nourished affluent society, including high blood pressure, fatty liver disease, increased rates of inflammation, and insulin resistance.[1]

Just think of the implications of these findings, at least possibly, if it turns out they can be replicated in humans. (Of course, this was not discovered using only two mice, but many. The research was carried out at the renowned Salk Institute for Biological Studies in San Diego, California, and the results published in top-level scientific journals such as *Cell Metabolism*.) As we already know, the standard explanation for obesity is based on the principle of energy balance. How do people get fat? By eating more than they burn. Either they eat too much or they exercise too little. And that in turn is based on the idea that a calorie is always a calorie, irrespective of *when* it is eaten. This is in tune with the first thought that goes through our head when we see an overweight person (or mouse!): wow, they must really shovel in the food (and be really lazy)!

'Energy balance' sounds like it's based on a perfectly logical law of physics, and, at a base level, that's true: one of the rules of our universe is that energy can't just disappear. That's also true of the energy we put inside our bodies. And yet, that principle falls short as soon as we add complex biological organisms into the mix — like mice, for example, or human beings.

Let's start by recalling the definition of a calorie. The calorie content of any food can be measured by burning samples of it, literally. Take a piece of carrot, for example, place it in a steel container filled with pure oxygen under pressure. Now, you just need to set the piece of carrot alight using electrodes — with a tiny stroke of lightning, so to speak. The steel container is submerged in water, and the rise in the temperature of that water is then measured as the carrot burns. The more the temperature rises, the more energy is contained in the food sample, so the higher its calorific value is. The kind of calorie we care about in nutrition science is specifically a 'kilogram calorie' or 'kilocalorie (kcal)'. A kilocalorie is nothing more than the amount of energy it takes to heat one kilogram of water by one degree Celsius.

So far, so good. But, to my knowledge, most steel containers don't care what time of day it is when they are 'fed' a food sample. For a steel container, a calorie really is always a calorie. But for living organisms, which have adapted to the rhythm of night and day as the earth rotated over millions of years, that's probably not the case. For such organisms, it may be a question not only of how many calories they eat, but also of *when* they eat.

Scientists have discovered in recent years just how important 'when' is. Our metabolism runs at completely different speeds depending on the time of day or night. I don't mean to say that our bodies defy the fundamental laws of physics — they really can't do that. But: calories are processed by our bodies differently, depending on when or in what rhythm or in what time window we consume them. For instance, by burning them and turning them into heat under certain circumstances, rather than storing them as fat deposits.[2] This isn't our body's biology defying physics, it's just introducing another layer of complexity to the equation. Who'd have thought it, but it turns out we are a little bit more complex than a steel container ...

Our body's day-and-night rhythm can be traced all the way to the inside of our cells, to our genes. More than half our genetic activity works on a day–night cycle.[3] That means thousands of our genes are sometimes more active and sometimes less, depending on the time of day or night. This cycle is called the 'circadian rhythm'. In certain organs, for example the liver, many genes are fired up in the early morning, while others are shut down. Depending on the time of day, the various cells of our organs produce different proteins due to the different gene activity. You could say we are a different organism at different times of the day, a different person, even — which I can definitely confirm as someone who is chronically grumpy in the mornings.

But isn't this all just theoretical hairsplitting? No, because the real-life consequences of all this are quite considerable. They can make the difference between your being fat or thin. If you give two test subjects identical meals, but give the meals to one of them in the morning and the other in the evening, their bodies will react completely differently, even if the fasting time between their meals is the same length. Our sensitivity to insulin is highest in the morning, for example, so the blood-sugar spike following a meal is least pronounced in the morning. That means we are better able to 'put away' food, especially carbohydrates, in the mornings. As the day goes on, we become increasingly less able to control our blood sugar. Purely from the point of view of our blood sugar, when we eat a meal in the evening, it's as if we were consuming a meal *twice the size* of one eaten earlier in the day, even if the actual size of the meals was identical. It's almost as if we turn into temporary diabetes patients in the late evening. And carbohydrates are then a particular problem.[4]

So what we eat when really does make a difference. In one experiment, scientists divided overweight women into two groups. All the women were asked to follow the same diet with the same (reduced) number of calories. The only difference between the two groups was that one ate a big breakfast and a small evening meal, while the other did exactly the opposite (small breakfast, big dinner). The result was that the group that ate the large breakfast lost far more weight. In addition, and probably in connection with the timing difference, their blood-fat levels were far better at the end of the experiment.[5] That doesn't mean we all have to force a monster breakfast down our necks, especially those of us who happen to belong to that part of the species that don't feel hungry just after waking up. Nonetheless, it's worth noting that it's better in general to take in most of your calories earlier in the day rather than later.[6]

One reason for the difficulty dealing with carbohydrates in the evening is the sleep hormone melatonin, the secretion of which is highly regulated by the circadian rhythm. Bright daylight suppresses the formation of melatonin. When it gets dark, the concentration of melatonin in our blood rises and we feel tired. The pancreatic cells that produce insulin are also equipped with receptor molecules for melatonin. When melatonin binds with those receptors, the secretion of insulin is inhibited.[7] The pancreas goes to sleep, in a way. This results in reduced blood-sugar regulation in the late evening and during the night. If we gorge on a pile of potatoes late in the evening, when our pancreas is already dozing, the weaker production of insulin means the glucose molecules will remain in the bloodstream for longer than usual, increasing the risk that they will 'clog up' our bodies from the inside.

In view of all this, it seems sensible to schedule the eating of healthy carbohydrates, such as wholemeal bread, muesli, and fruit, for the first half of the day. In the afternoon, you can move on to a protein source, such as a fillet of fish with some salad and vegetables, while your evening meal should be more oriented towards foods that are higher in fat, such as avocados, nuts, olive oil, cheese ...[8]

Even more important than this 'fine tuning' of when to eat from each of the main food groups is limiting the time when you actually eat your meals to a particular part of the day. The only way to get this right is to test it out on yourself by trial and error. For me, personally, the best time window is 8.00 am to 8.00 pm. God knows I'm no master of self-discipline, but I find that it's usually not too difficult to keep to that 12-hour window. If I'm feeling particularly disciplined, or have a bit of extra fat I need to lose, I reduce the window to between 9.00 am and 7.00 pm. Although it's not possible at the moment to say from a scientific point of view what window is the best, mouse experiments indicate that a good rule of thumb is that the smaller the window, the greater the health effects. Ultimately, though, what's most important is to find a rhythm that suits you and your lifestyle. Find a rhythm that doesn't feel too much like a sacrifice!

Now let's take a brief look at what effect time-restricted eating has on the body — perhaps it will motivate you to start testing out possible time windows, if the above hasn't convinced you already. Why is time-restricted eating good for us at all? Shouldn't it be better from our body's point of view for it to receive a *constant* supply of precious energy and valuable nutrients?

Effect no. 1: time-restricted eating stabilises the circadian rhythm

One advantage of limiting your eating to a certain period of the day is that it helps the body keep to the natural, light-controlled day–night cycle. As sure as day follows night, all the organs of our body are connected to this rhythm, all the way down to the genetic level. Put simply, we could say that the gene activity of our organs — gut, liver, pancreas, etc. — means they expect a meal in the morning. In fact, they are (ener)genetically hungry.

As the day goes on, the pattern of gene activity in our body changes. The cells in our bodies switch to a different mode of action. Just like us, our cells can't do everything all at once. The night-time, when our cells are not being bombarded with nutrients that require processing, is a good time for some cells to get some tidying up done, for example. Agglutinated or otherwise-damaging protein structures and defective organelles can now be broken down in peace. It's comparable, in a way, to a street party, where the festivities would be ruined if the sanitation department turned up and started clearing away before the partying was finished. But the street sweepers come at night when the party's over. Our cells act this way — if we give them the chance.

If, on the other hand, we raid the fridge at night (not an uncommon occurrence for me not so long ago) to wolf down some scoops of double-choc-chip ice cream, our cells can forget any ideas of clearing up. Genes in the liver and other organs that were looking forward to a well-earned rest are 'violently' shaken awake by the unexpected storm of calories that now need processing. mTOR gets into gear. Other genes, including ones responsible for triggering the clear-up and repair work inside the cell, are silenced by the onslaught. The usually harmonious rhythm of gene activity is destroyed by our night-time snack attack.

By the same token, consistently eating within a limited time window stabilises the body's day-and-night rhythm, which has been proven to improve sleep.[9] This rhythm often becomes weaker as we get older; sleep patterns are less fixed, unravelling and causing sleep problems. A strict daytime eating window — and no moonlit expeditions to the fridge — can be helpful in combating this. In short, our bodies operate better when the rhythms of darkness and light and eating and fasting are in harmony with each other.[10]

Effect no. 2: big and little fasts are good for you

The second reason why time-restricted eating is good for our body is the breaks without eating themselves. All humanity, according to the cliché at least, used to be satisfied with the classic three meals a day. We modern urban dwellers, however, spend the entire day eating, snacking, and grazing — not even sunset slows us down.

This uninterrupted flood of food and energy — together with the encouragement of insulin, IGF-1, and mTOR — places our cells in permanent growth mode, which basically means our cells are continuously ageing. However, if we stop eating for a while, our insulin and IGF-1 levels sink and mTOR comes to rest as well. The cells launch their autophagy program; our bodies have changed from growth mode to clean-up mode.

In this way, we go on a short fast every night as our bodies are maintained and 'overhauled'. (Scientists have recently discovered that the brain cleans itself at night, too. This process can even flush out some of the protein aggregates suspected of causing Alzheimer's disease!)[11] The word 'breakfast' actually says it all — we fast through the night, and break that fast in the morning.

There's currently some hype surrounding the idea of fasting, which is perhaps not all that surprising in this time of unprecedented abundance of food. This excess is what makes voluntary sacrifice so attractive to us. It says something about us as people. About our character, our capacity for self-discipline. These days, we can afford, if you like, to revel in fasting.

I welcome this development, by and large. But I also think we overestimate the benefits of our classic idea of fasting, at least as far as the physical effects on our bodies are concerned. By 'classic idea of fasting', I mean the radical act of self-starvation or near starvation for a few days once or twice a year. Followed by a return to 'business as usual'.

It's not that I think such fasts are pure nonsense: I have experimented with fasting myself and found the experience extremely enriching. But I find fasting for several days difficult. Going to bed hungry sounds so simple. It's easy to write the phrase. Right, I'll go to bed hungry! But when the moment actually arrives, it's bloody hard. The good thing about it is that, when you wake up the next morning, you realise you're still alive and everything's basically more-or-less okay. That's definitely the most important lesson I learned from my experiments with fasting: to experience doing without something you thought was absolutely essential.

The following trick helps to make fasting a little bit easier: several days *before* your fast begins, switch to a low-carb, high-fat diet. When the body lacks carbohydrates, it has to rely more on its fat-burning mode. Interestingly enough, this is similar to fasting itself, when the body has no choice but to burn fat. During fasting, the body's carbohydrate stores (glycogen) are quickly used up, just like when we are on a reduced-carb diet. Our body then has to dig into its fat reserves. The machinery of our bodies is then fuelled more by fat rather than glucose. From the point of view of our cells, it probably makes little difference whether the fat comes from our food or our own body's stores.[12]

Not eating for a couple of days also helps to recalibrate our appreciation of food. A simple strawberry becomes a taste explosion once eating them is finally allowed again. Overall, many people report that fasting can be an inspiring, consciousness-expanding experience. Remembering my own fasting experience helps me when I find myself in a situation where the only food available all around is junk food. Fine — I just won't eat anything. I know I can cope. A temporary bout of hunger is fine.

Fasting has been thought for centuries to have a therapeutic effect on many physical conditions, and scientific studies in recent years have corroborated this old folk wisdom. Perhaps the most important and best-documented example is that of type-2 diabetes. One of the problems, perhaps the central problem, of type-2 diabetes is that the inner organs, such as the liver, become fatty, as do the muscles, which then become numb to the signals from insulin telling them to take up glucose from the bloodstream.

As mentioned at the start of this book, researchers at the University of Newcastle in the United Kingdom placed a group of overweight diabetics on a highly calorie-reduced diet for six weeks. All they were given was a solution of nutrients and three modest portions of vegetables. The subjects were also encouraged to drink at least two litres of water per day. This wasn't quite full-on water fasting, but the subjects' total energy intake was only 600 calories per day (compared to their usual 2,000 or more).

The effects were phenomenal. Even after just one week, the accumulated fat in the patients' livers was reduced by 30 per cent and their liver cells had started responding to insulin once again. Their fasting blood-sugar levels normalised rapidly. Gradually, the accumulated fat disappeared from their pancreas — the organ that synthesises insulin. At the end of their eight-week diet, the patients' reaction to insulin had returned to the level of a healthy person. (Note: if you plan to go on a fast of this kind, you should under no circumstances attempt it without the supervision of your doctor, especially if you regularly take medication, which will need to be dose-adjusted. A fasting cure is so efficient that you will probably require less or no medication if you are a diabetes patient!)[13]

Fasting has been found to have similarly positive effects on high blood pressure[14] and rheumatism.[15] These results are all very impressive; however, they are extreme examples, and a drastic fasting cure can ultimately 'only' be the start of what will hopefully be a life-long change in eating habits.

After all, what's ultimately important for our physical wellbeing is not something we do once or twice a year, but what we do every day. Take physical exercise as a comparison. At the very basic level, sporting activity creates an energy deficit, just like fasting. Also, just like fasting, exercise leads to an increase in insulin sensitivity and a decrease in blood pressure. Still, few people would consider it a good idea, or particularly beneficial to our long-term health, to complete an ultra-intensive five-day sports program once or twice a year and not do any exercise otherwise. Rather, we know that *regular* exercise is what does us good. That's why our classic idea of fasting is overrated. The power of fasting is most efficiently harnessed when we manage to build 'small fasts' into our daily routine, in the form of time-restricted eating, for example. I admit that eating nothing after 8.00 pm every evening is not going to transport you to the heights of ecstasy. It's not going be a transcendental experience. This 'little' form of fasting is not a path to enlightenment, just as the experience of going for a daily run pales in comparison to the challenges and highlights of running a marathon. Of course, the two aren't mutually exclusive. But when it comes to ageing healthily, I think a daily run is probably more important.

Only eating within a restricted window of time is not the only way mini-fasting can be integrated into a daily routine. Those with more self-discipline than me may find it helpful to go without eating for a full day or even two days per week, or at least to eat significantly less.[16] This is probably also a good way to kickstart the autophagy process.

Some people shouldn't fast at all, not even for a couple of days. These include pregnant women and nursing mothers, as well as women who are trying to get pregnant (when a woman fasts, she sends her body a signal that there's a period of food shortage at the moment, which is naturally not a good time for her body to be making a baby). Children should be growing, not fasting. Elderly and underweight people should also avoid fasting, among other reasons because an extended hunger diet leads not only to fat being broken down, but also to muscle tissue.

The (softer) option of time-restricted eating, by contrast, is probably something that would do all of us good, not least because it supports the natural rhythm inside all of us (except babies, naturally). For some people, time-restricted eating might seem obvious because it's what they naturally do anyway, but, for many these days, this is no longer the case. Until new research is released to prove otherwise, I'll continue to consider time-restricted eating to be the easiest and most effective way of fasting.

Conclusion

When it comes to the issues of fasting and when to eat, there's no shortage of misconceptions and half-truths out there. One very persistent myth is that breakfast is the most important meal of the day and should not be missed under any circumstances. Some people consider prolonged fasting to be harmful, others think it's a miracle cure. Taking a sober view, the following conclusions can be drawn:

- If you don't feel hungry in the morning, there's no need to force yourself to eat breakfast, just because it is supposed to be 'the most important meal of the day'. In fact, this can be an opportunity to extend your night-time fast for a little longer. By and large, however, it does seem to be better to take in most of your daily calories in the first half of the day, or at least not shortly before bedtime.

- Our sensitivity to insulin is highest in the morning. That means our body is best able to cope with rapidly digested, high-carb food at that time of the day. Our bodies become increasingly resistant to insulin as the day goes on, which means if you're going to eat potato gratin or a pile of pasta, it's best to do it at lunchtime rather than suppertime.

- It can be beneficial (in the case of obesity and geriatric conditions) to limit mealtimes to within a certain period of time during the day, for example from 8.00 am to 8.00 pm. It doesn't appear to be important — for weight or general health — whether you eat several small meals during that window or two or three big meals.[17] It's far more important to avoid eating a huge meal in the late evening, and not to eat at all through the night.

- Fasting for a number of days is not harmful. In fact, the opposite is the case: it is an effective way of kickstarting our cells' self-cleaning program. Positive effects have been proven in particular in the case of diabetes and rheumatism.

- The question of the most effective way to fast remains unanswered. Comparing fasting to other 'healthy' activities (exercising, healthy eating, sleeping, relaxing, etc.), it seems *regularity* is the key. Radically starving your body for a few days once a year probably has as much or as little long-term effect on our bodies as completing a one-week intensive exercise program once a year. Regular 'little' fasts (e.g. never eating at night, or fasting for 24 hours once a week) are probably more effective in the long run.

My 12 most important diet tips

1. Eat real food

The first and most important rule is to eat unprocessed foods at far as possible. That means anything that comes directly from nature, in its natural state. It's anything that doesn't need a list of ingredients on the packet. Any kind of vegetable, and any kind of fruit. Fish and meat in moderate amounts. Some people call this 'real food'. It's basically the produce you see first when you enter a supermarket. Or what you can buy at a traditional market.

Some food is processed, but still healthy. One example is wholemeal products like wholemeal bread or minimally processed rolled oats, yoghurt, and cheese. Olive oil (extra virgin), cold-pressed rapeseed oil, tea, and coffee are also in this category. As long as the recommended amounts aren't exceeded, I would also place wine and beer in this group. Usually, the processed foods that are nonetheless healthy are the ones with thousands of years of tradition behind them.

In practical terms, this rule could be rewritten like this: cook. Of course, cooking healthy meals for yourself from scratch using fresh ingredients takes time (but you will get that time back later by having a longer and healthier life). And, also of course, it's easier just to shove a readymade pizza in the oven. But shoving a fresh fish in the oven is hardly more difficult. In this regard, I find the mixed salads sold at most supermarkets these days very practical, as well as ready-prepared vegetables, although both tend to go off very quickly. (One of my super-simple favourites takes no more than a quarter of an hour to make: a piece of baked salmon garnished with rosemary on a bed of mixed salad with mixed seeds and olive-oil dressing. Or for lunch, a thick slice of wholegrain bread with smashed avocado, sometimes with a poached egg on top.)

If you often find yourself in places or situations where there's no real food to be had far and wide (at conferences, at train stations), don't make excuses ('It's not my fault! What choice do I have? I'm a victim of circumstance!'). Be smarter than the circumstance! At the start of the day, prepare a box with your favourite fruit, vegetables, or a wholemeal sandwich. Take an apple and a bag of nuts with you. Or just don't eat for a couple of hours. Don't spoil your precious appetite with junk food. Don't make any compromises when it comes to good, real food.

2. Make plants your main meal

The second most important rule is to eat more plant-based food and less that comes from animal sources. Vegetables shouldn't be an accompaniment to meat; it should be the other way around. Basically, all edible plants and mushrooms in their natural state are simply the healthiest things you can eat. It doesn't matter how you eat them — raw, boiled, or steamed. There's barely a plant-based food that you can eat too much of (important exceptions are potatoes and rice).

As soon as you move into the realm of *processed* plant-based foods, you'll soon come across unhealthy fare, and the extreme examples are of course refined sugar and white flour. Potato chips and fries are plant-based junk food. Sugar, white flour, fries, chips — these examples show that vegan food is not always the healthiest option, although a vegan diet *can* be very healthy. So what I mean here is plant-based foods that are still recognisable as belonging to the plant kingdom.

3. Eat fish rather than meat

We can draw up a clear hierarchy of healthiness when it comes to meat and fish. Oily fish and shellfish are the healthiest (deep-fried 'fishoids' don't count). Then comes white meat from chicken and turkey, especially if the animals the meat comes from led healthy lives themselves (for that reason if no other, I always avoid factory-farmed meat). If you love red meat like beef and pork, then just eat it occasionally, and in its unprocessed form. No sausages! No hotdogs! My rough personal rule of thumb is: fish once or twice a week, white meat once or twice a month, and a piece of grass-fed steak, wild game meat, or a free-range Sunday roast once or twice a year. My preferred 'protein source' is pulses such as lentils, beans, and chickpeas (as well as bulgur, nuts, linseeds, chia seeds, and wheatgerm).

4. Yoghurt: yes. Cheese: okay, too. Milk: not so much

The crucial question when dealing with dairy products is not whether they are low-fat or full-fat, but whether they are *fermented* or not. Yoghurt (or alternatively, kefir) is particularly recommended, due to its slimming properties. Cheese is okay. I consider cow's milk to be relatively unsuitable for adults: to be on the safe side, I would drink no more than one to two glasses a day (myself, I only use milk in my coffee). Yoghurt can easily be combined with all kinds of delicious things to make it taste even better. I eat a bowl of yoghurt every day, with blueberries or strawberries. If you want, you can also try it with wheatgerm, linseeds or chia seeds, nuts, or some porridge oats. And just a bit of grated dark chocolate on top to finish it off …

5. Reduce sugar to a minimum, avoid industrial trans fats completely

Reducing sugar to a minimum doesn't mean avoiding it at all costs. A jar of red cabbage or beetroot containing a bit of added sugar, or a little sugar sprinkled on your muesli is often still a good choice compared to what you would otherwise eat in their place. Some foods, for example wheatgerm, naturally contain some sugar, but wheatgerm is otherwise so chock-full of beneficial nutrients (plant protein, fibre, vitamin E, folic acid, omega-3 fatty acids, spermidine …) that I eat a spoonful every day. Avoid industrially produced snacks like potato chips, biscuits, and cookies, and anything in your local bakery that attracts the wasps.

6. Don't be afraid of fat!

Fat doesn't make you fat per se. Ironically, it's especially good for obese people to eat more healthy fat (remember: insulin resistance). Polyunsaturated and monounsaturated fatty acids are the best fats to eat. In actual food terms, that means you can enjoy nuts of any kind, but especially those you find most tasty (rule of thumb: two handfuls of nuts a day; personally, I eat nuts with practically every other meal, and between meals, and, in fact, all the time). Eat oily fish such as salmon and herring, as well as the much-mentioned linseeds and chia seeds, as well as sunflower and other types of seeds. Other excellent sources of fat include avocados, olive oil, and rapeseed oil. Also, as mentioned often, cheese is also a good choice. Butter is okay in moderation.

7. Weight-loss tip no. 1: low-carb is not just a 'fad diet', but rather definitely worth a try, especially for people with obesity

On average, low-carb diets appear to be quite an efficient way to lose weight. It depends on every individual body, so trying things out for yourself is the best method. However, particularly if you have insulin resistance — a common consequence of obesity — you should avoid rapidly digested, high-carb foods like white bread, potatoes, and rice (and certainly sugar and fruit juice, but *not* whole fruit or 'slow carbs', such as the blessed pulses, of course). It's important to remember that a low-carb diet doesn't necessarily mean the Atkins diet. At the end of chapter 5 (pages 124–126), I outlined the components of a healthy low-carb diet. Those who want to get their weight back under control should experiment with this kind of diet for at least two to three weeks to find out how their body reacts.

8. Weight-loss tip no. 2: make use of the protein-leverage effect

As far as their ability to make us feel full is concerned, one calorie is not always the same as another. Protein is far better at making us feel full than fat or carbohydrates. If you want to lose weight, try to smuggle a bit more protein into your diet — for example, in the form of yoghurt, fish and shellfish, nuts, seeds, and, in particular, all kinds of pulses (the many different kinds of beans, peas, chickpeas, lentils). Eggs in moderation — rule of thumb: a maximum of one egg a day on average.

9. Weight-loss tip no. 3: practise 'time-restricted eating'

I think a very simple way to maintain your weight is to eat only during a limited time window, for example between 8.00 am and 8.00 pm (the '8 till 8' rule). Up to a point, my advice is: the smaller the window, the greater the effect. No night-time trips to the fridge! You're not hungry when you wake up in the morning? Great, listen to your body, skip breakfast, and stretch out your night fast a little longer. Still, it's best to consume most of your daily calories in the first half of the day (rather than eating a monster meal late in the evening). Personally, I still often allow myself a hearty evening meal, but stop eating at least two, usually three or four hours before going to bed. A glass of calorie-free water is always okay, of course.

10. Weight-loss tip no. 4: reduce brain inflammation with omega-3

Obesity can be associated with an inflammation in that part of the brain that's responsible for our feeling of fullness after eating (hypothalamus). It's as if the brain had a head cold. The hypothalamus can no longer 'smell' the signals coming from the body to say that it's full. The result is that we feel hungry, although, or precisely because, we are overweight. Omega-3 fatty acids have an anti-inflammatory effect and so can also be helpful in losing weight. The brain's 'head cold' gets better, the satiety centre in the brain can once again react to the body's signals, and we no longer feel hungry. Good sources of omega-3 include walnuts, chia seeds and linseeds, rapeseed oil, and, especially, oily fish. Alternatively, as a second choice, omega-3 capsules (fish, krill, or algae oil).

11. No vitamin pills!

Even that shouldn't be taken dogmatically. The most important exception is vitamin D_3 (1,000 to 2,000 international units a day). Perhaps also omega-3 and a vitamin-B complex. Vegetarians, and especially vegans, are advised to take at least a vitamin-B_{12} supplement.

The B vitamin folic acid is also an exception — we consume too little of it on average and would profit from consuming more[1] (this is all the more true for those who regularly drink alcohol). The healthiest sources of folic acid are Brussels sprouts, cos/romaine lettuce (which is delicious), boiled spinach, asparagus, pulses, wheatgerm, broccoli, avocados, and oranges.

A note on salt: use it sparingly,[2] and choose iodised salt. Experiment with using more herbs for flavour, such as rosemary, thyme, parsley, or more exotic flavours like cinnamon, turmeric, etc. Even a squirt of lemon juice can pep up a meal no end!

12. Enjoy!

Many might say: this is all well and good, but what about enjoyment? Isn't it a bit sad, dear Mr Kast, to look only at the health aspect of what we eat? What is this whole 'nutrition cult' anyway? Well, if you ask me in such a personal way, I can only answer personally, and I can honestly say this: quite apart from the fact that it is extremely satisfying to feel that I am fit, and free of any heart complaints (I really can't put into words how much I enjoy and appreciate that feeling), I now enjoy my food more than I ever did before. Did I use to enjoy my potato chips and fries? Yes, in a way, I did. Today, I'm no longer tempted by all that junk food (notwithstanding my grandma's delicious *dampfnudeln*). Ultimately, everyone has to find their own path to enjoyment and good health. I don't think the two are mutually exclusive. Certainly not for me. I have never had the feeling that I've fallen victim to any kind of nutrition cult. Dogmas are just not my thing. I (almost) never feel my new nutritional way of life is a sacrifice — on the contrary, it's a culinary enrichment for my life. It's pure, simple, and often simply delicious.

Bibliography

Ables et al. (2016): *Annals of the New York Academy of Sciences*, 1363, pp. 68–79

Aeberli et al. (2011): *American Journal of Clinical Nutrition*, 94, pp. 479–85

Aeberli et al. (2013): *Diabetes Care*, 36, pp. 150–6

Ahmed et al. (2013): *Current Opinion in Clinical Nutrition and Metabolic Care*, 16, pp. 434–9

Alhassan et al. (2008): *International Journal of Obesity*, 32, pp. 985–91

Ali et al. (2011): *Food & Nutrition Research*, 55, p. 5572

Ames (2005): *EMBO Reports*, 6, pp. 20–4

Andersen et al. (2012): *Journal of Gerontology: Biological Sciences*, 67A(4), pp. 395–405

Anderson et al. (2016): *Journal of the American Heart Association*, 5, p. e003815

Appel & Van Horn (2013): *NEJM*, 368, pp. 1353–4

Atkins (1999): *Die neue Atkins-Diät*. Goldmann

Atkinson et al. (2008): *Diabetes Care*, 31, pp. 2281–3

Aune et al. (2016): *British Medical Journal*, 353, p. i2716

Bagnardi et al. (2008): *Journal of Epidemiology and Community Health*, 62, pp. 615–19

Bagnardi et al. (2014): *British Journal of Cancer*, 112, pp. 580–93

Bao et al. (2009): *Journal of Clinical Nutrition*, 90, pp. 986–92

de Batlle et al. (2015): *Journal of the National Cancer Institute*, 107, p. dju367

Bayless et al. (2017): *Current Gastroenterology Reports*, 19, p. 23

Beauchamp et al. (2005): *Nature*, 437, p. 45

Belin et al. (2011): *Circulation Heart Failure*, 4, pp. 404–13

Béliveau & Gingras (2007): *Krebszellen mögen keine Himbeeren*. Kösel

Bell et al. (2014): *American Journal of Epidemiology*, 179, pp. 710–20

Bell et al. (2017): *British Medical Journal*, 356, p. j909

Bellavia et al. (2014): *Annals of Epidemiology*, 24, pp. 291–6

Bellavia et al. (2017): *Journal of Internal Medicine*, 281, pp. 86–95

Bender et al. (2014): *Obesity Review*, 15, pp. 657–65

Bettuzzi et al. (2006): *Cancer Research*, 66, pp. 1234–40

Bjelakovic et al. (2014): *Cochrane Database of Systematic Reviews*, online 10 January

Blagosklonny (2009): *Cell Cycle*, 8, pp. 4055–9

Blundell et al. (2015): *Obesity Reviews*, 16, pp. 67–76

Brand-Miller et al. (2009): *American Journal of Clinical Nutrition*, 89, p. 97–105

Brand-Miller et al. (2010): *The Low GI Handbook*. Da Capo Press

Bredesen (2014): *Aging*, 6, pp. 707–17

Bredesen (2017): *The End of Alzheimer's*. Vermilion

Bredesen et al. (2016): *Aging*, 8, p. 1–9

Brien et al. (2011): *British Medical Journal*, 342, p. d636

Buettner (2015): *The Blue Zones Solution*. National Geographic Society

Burr et al. (1989): *Lancet*, 334, pp. 757–61

de Cabo et al. (2014): *Cell*, 157, pp. 1515– 26

Cai et al. (2012). *European Journal of Clinical Nutrition*, 66, pp. 872–7

Calder (2015): *Journal of Parenteral and Enteral Nutrition*, 39, pp. 18S–32S

Calder (2016): *Annals of Nutrition & Metabolism*, 69, pp. 8–21

Cantley (2014): *BMC Biology*, 12, p. 8

Cao et al. (2015): *British Medical Journal*, 351, p. h4238

Cardoso et al. (2016): *Nutrition Research Reviews*, 29, pp. 281–94

Carey et al. (2015): *PLOS One*, 10, p. e0131608

Casal et al. (2010): *Food and Chemical Toxicology*, 48, pp. 2972–9

Catenacci & Wyatt (2007): *Nature Clinical Practice Endocrinology & Metabolism*, 3, pp. 518–29

Caudwell et al. (2009): *Public Health Nutrition*, 12, pp. 1663–6

Cavuoto & Fenech (2012): *Cancer Treatment Reviews*, 38, pp. 726–36

Chaix et al. (2014): *Cell Metabolism*, 20, pp. 991–1005

Chen et al. (2014): *British Journal of Cancer*, 110, pp. 2327–38

Chen et al. (2016): *Scientific Reports*, 6, p. 28165

Chhetry et al. (2016): *Journal of Psychiatric Research*, 75, pp. 65–74

Chin et al. (2016): *Obesity Reviews*, 17, pp. 1226–44

Chowdhury et al. (2014): *British Medical Journal*, 348, p. g1903

Chuengsamarn et al. (2012): *Diabetes Care*, 35, pp. 2121–7

Cintra et al. (2012): *PLOS One*, 7, p. e30571

Cladis et al. (2014): *Lipids*, 49, pp. 1005–18

Clifton et al. (2014): *Nutrition, Metabolism & Cardiovascular Diseases*, 24, pp. 224–35

Costanzo et al. (2011): *European Journal of Epidemiology*, 26, pp. 833–50

Costello et al. (2016): *Journal of the Academy of Nutrition and Dietetics*, online 8 September

Couzin-Frankel (2014): *Science*, 343, p. 1068

Daley et al. (2010): *Nutrition Journal*, 9, p. 10

Dansinger et al. (2005): *JAMA*, 293, pp. 43–53

Darmadi-Blackberry et al. (2004): *Asia Pacific Journal of Clinical Nutrition*, 13, pp. 217–20

Davis (2013): *Weizenwampe*. Goldmann

Dehghan et al. (2017): *Lancet*, online 29 August

Dennison et al. (2017): *Nature Reviews Rheumatology*, 13, pp. 340–7

Desai et al. (2016): *Cell*, 167, pp. 1339–53

Di Castelnuovo et al. (2006): *Archives of Internal Medicine*, 166, pp. 2437–45

Douaud et al. (2013): *PNAS*, 110, pp. 9523–8

Due et al. (2004): *International Journal of Obesity*, 28, pp. 1283–90

Eenfeldt (2013): *Echt fett*. Ennsthaler Verlag Steyr

Eisenberg et al. (2009): *Nature Cell Biology*, 11, pp. 1305–14

Eisenberg et al. (2016): *Nature Medicine*, online 14 November

Esatbeyoglu et al. (2016): *Journal of Agricultural and Food Chemistry*, 64, pp. 2105–11

Escarpa & González (1998): *Journal of Chromatography A*, 823, pp. 331–7

Esposito et al. (2009): *Annals of Internal Medicine*, 151, pp. 306–14

Esselstyn (2001): *Preventive Cardiology*, 4, pp. 171–7

Esselstyn (2015): *Essen gegen Herzinfarkt*. Trias Verlag

Esselstyn et al. (2014): *Journal of Family Practice*, 63, pp. 356–64

Estruch et al. (2013): *NEJM*, 368, pp. 1279–90

Eyres et al. (2016): *Nutrition Reviews*, 74, pp. 267–80

Fardet (2010): *Nutrition Research Reviews*, 23, pp. 65–134

Fardet (2015): *Food & Function*, 6, pp. 363–82

Fardet & Boirie (2013): *Nutrition Reviews*, 71, pp. 643–56

Fardet & Boirie (2014): *Nutrition Reviews*, 72, pp. 741–62

Farin et al. (2006): *American Journal of Clinical Nutrition*, 83, pp. 47–51

Fetissov (2017): *Nature Reviews Endocrinology*, 13, pp. 11–25

Finkel (2015): *Nature Medicine*, 21, pp. 1416–23

Folkman & Kalluri (2004): *Nature*, 427, p. 787

Fontana et al. (2010): *Science*, 328, pp. 321–6

Fontana & Partridge (2015): *Cell*, 161, pp. 106–18

Fraser & Shavlik (2001): *Archives of Internal Medicine*, 161, pp. 1645–52

Freedman et al. (2012): *NEJM*, 366, pp. 1891–1904

Fries (1980): *NEJM*, 303, pp. 130–5

Fries et al. (2011): *Journal of Aging Research*, online 23 August

Fry et al. (2011): *Skeletal Muscle*, 1, p. 11

Furman et al. (2017): *Nature Medicine*, 23, pp. 174–84

Gardner (2012): *International Journal of Obesity Supplements*, 2, pp. S11–S15

Gardner et al. (2007): *JAMA*, 297, pp. 969–77

Gea et al. (2014): *British Journal of Nutrition*, 111, pp. 1871–80

Gepner et al. (2015): *Annals of Internal Medicine*, 163, pp. 569–79

Ghorbani et al. (2014): *International Journal of Endocrinology and Metabolism*, 12, p. e18081

Gil & Gil (2015): *British Journal of Nutrition*, 113, pp. S58–S67

Gill & Panda (2015): *Cell Metabolism*, 22, pp. 789–98

Giuseppe et al. (2014a): *Arthritis Research & Therapy*, 16, p. 446

Giuseppe et al. (2014b): *Annals of the Rheumatic Diseases*, 73, pp. 1949–53

del Gobbo et al. (2016): *JAMA Internal Medicine*, 176, pp. 1155–66 de Goede et al. (2015): *Nutrition Reviews*, 73, pp. 259–75

Goldhamer et al. (2002): *Journal of Alternative and Complementary Medicine*, 8, pp. 643–50

Goletzke et al. (2016): *European Journal of Clinical Nutrition*, online 2 March

Gosby et al. (2011): *PLOS One*, 6, p. e25929

Gosby et al. (2014): *Obesity Reviews*, 15, pp. 183–91

Grassi et al. (2005): *American Journal of Clinical Nutrition*, 81, pp. 611–14

Grassi et al. (2008): *Journal of Nutrition*, 138, pp. 1671–6

Graudal et al. (2014): *American Journal of Hypertension*, 27, pp. 1129–37

Green et al. (2017): *Nature Reviews Disease Primers*, 3, p. 17040

Greger (2015): *How Not to Die*. Flatiron

Grosso et al. (2015): *Critical Reviews in Food Science and Nutrition*, online 3 November

Grosso et al. (2017): *Annual Review of Nutrition*, 37, pp. 131–56

Gu et al. (2015): *Neurology*, 85, pp. 1–8

Guasch-Ferré et al. (2017): *Journal of the American College of Cardiology*, 70, pp. 2519–32

Guimarães et al. (2015): *Food Science & Nutrition*, 4, pp. 398–408

Gunter et al. (2017): *Annals of Internal Medicine*, online 11 July

Guo et al. (2017a): *European Journal of Epidemiology*, 32, pp. 269–87

Guo et al. (2017b): *Medicine*, 96, p. e6426

Hadjivassiliou et al. (2014): *Handbook of Clinical Neurology*, 120, pp. 607–19

Halsted et al. (2002): *Journal of Nutrition*, 132, pp. 2367S–2372S

Han et al. (2007): *Metabolism Clinical and Experimental*, 56, pp. 985–91

Harrison et al. (2009): *Nature*, 460, pp. 392–5

Harrison et al. (2017): *Cancer Causes & Control*, 28, pp. 497–528

Hatori et al. (2012): *Cell Metabolism*, 15, pp. 848–60

Henriques et al. (2014): *British Journal of Nutrition*, 112, pp. 964–75

Hermsdorff et al. (2011): *European Journal of Nutrition*, 50, pp. 61–9

Hibbeln (2002): *Journal of Affective Disorders*, 69, pp. 15–29

Hjorth et al. (2017): *American Journal of Clinical Nutrition*, online 5 July

Hoffman & Gerber (2014): *British Journal of Nutrition*, 112, pp. 1882–95

Hojsak et al. (2015): *Journal of Pediatric Gastroenterology and Nutrition*, 60, pp. 142–5

Holick (2017): *Reviews in Endocrine and Metabolic Disorders*, 18, pp. 153–65

Holick et al. (2011): *Journal of Clinical Endocrinology & Metabolism*, 96, pp. 1911–30

Holländer et al. (2015): *American Journal of Clinical Nutrition*, 102, pp. 556–72

Holst et al. (2017): *Diabetologia*, online 27 July

Hosios et al. (2016): *Developmental Cell*, 36, pp. 540–9

Howitz & Sinclair (2008): *Cell*, 133, pp. 387–91

Huang et al. (2012): *Clinical Nutrition*, 31, pp. 448–54

Hutchison et al. (2017): *Nutrients*, 9, p. 222

Jacka et al. (2017): *BMC Medicine*, 15, p. 23

Jacques & Wang (2014): *American Journal of Clinical Nutrition*, 99, pp. 1229–34

Jais & Brüning (2017): *Journal of Clinical Investigation*, 127, pp. 24–32

Jakubowicz et al. (2013): *Obesity*, 21, pp. 2504–12

Jenkins et al. (2012): *Archives of Internal Medicine*, 172, pp. 1653–60

Jernerén et al. (2015): *American Journal of Clinical Nutrition*, pp. 215–21

Ji et al. (2014): *British Journal of Cancer*, 112, pp. 149–52

Johnsen et al. (2015): *British Journal of Nutrition*, 114, pp. 608–23

Johnson (2015): *Der Fettschalter. Fettleibigkeit neu denken, verstehen und bekämpfen.* Hachinger

Johnson et al. (2013): *Nature*, 493, pp. 338–45

Johnston et al. (2014): *JAMA*, 312, pp. 923–33

Joven et al. (2014): *Critical Reviews in Food Science and Nutrition*, 54, pp. 985–1001

Juanola-Falgarona et al. (2014): *Journal of Nutrition*, 144, pp. 743–50

Kahleova et al. (2017): *Journal of Nutrition*, online 12 July

Kamiloglu et al. (2014): *Journal of the Science of Food and Agriculture*, 94, pp. 2225–33

Kanarek & Ho (1984): *Physiology & Behavior*, 32, pp. 639–45

Kaplan et al. (2017): *Lancet*, online 17 March

Karagas et al. (2016): *JAMA Pediatrics*, 170, pp. 609–16

Kavanagh et al. (2007): *Obesity*, 15, pp. 1675–84

Kennedy (2016): *Nutrients*, 8, p. 68

Kennedy & Lamming (2016): *Cell Metabolism*, 23, pp. 990–1003

Kessler et al. (2017): *Scientific Reports*, 7, p. 44170

Khanfar et al. (2015): *Phytotherapy Research*, 29, pp. 1776–82

Kim et al. (2016): *American Journal of Clinical Nutrition*, 103, pp. 1213–23

Knott et al. (2015): *British Journal of Medicine*, 350, p. h384

Knowler et al. (2002): *NEJM*, 346, pp. 393–403

Koh et al. (2016): *Cell*, 165, pp. 1332–45

Konner & Eaton (2010): *Nutrition in Clinical Practice*, 25, pp. 594–602

Kristensen et al. (2016): *Food & Nutrition Research*, 60, p. 32634

Laakso & Kuusisto (2014): *Nature Reviews Endocrinology*, 10, pp. 293–302

Lagiou et al. (2012): *British Medical Journal*, 344, p. e4026

Laplante & Sabatini (2012): *Cell*, 149, pp. 274–93

Larsson & Orsini (2014): *American Journal of Epidemiology*, 179, pp. 282–9

Latreille et al. (2012): *PLOS One*, 7, p. e44490

Lebwohl et al. (2015): *British Medical Journal*, 351, h4347

Lee et al. (2008): *PNAS*, 105, pp. 2498–2503

Lee et al. (2015a): *Journal of Microbiology and Biotechnology*, 25, pp. 2160–8

Lee et al. (2015b): *European Journal of Clinical Nutrition*, 69, pp. 1048–52

Leidy et al. (2015): *American Journal of Clinical Nutrition*, 101, pp. 1320S–1329S

Lesser et al. (2007): *PLOS Medicine*, 4, p. e5

Leung et al. (2014): *American Journal of Public Health*, 104, pp. 2425–31

Levine et al. (2014): *Cell Metabolism*, 19, pp. 407–17

Levkovich et al. (2013): *PLOS One*, 8, p. e53867

Li et al. (2012): *Heart*, 98, pp. 920–5

Lim et al. (2011): *Diabetologia*, 54, pp. 2506–14

Liu et al. (2013): *American Journal of Clinical Nutrition*, 98, pp. 340–48

Longo & Panda (2016): *Cell Metabolism*, 23, pp. 1048–59

de Lorgeril et al. (1994): *Lancet*, 343, pp. 1454–59

Luciano et al. (2017): *Neurology*, 88, pp. 449–55

Lustig (2016): *Die bittere Wahrheit über Zucker*. Riva

Lyssiotis & Cantley (2013): *Nature*, 502, pp. 181–82

Maddocks et al. (2017): *Nature*, 544, pp. 372–6

Madeo et al. (2015): *Journal of Clinical Investigation*, 125, pp. 85–93

Maersk et al. (2012): *American Journal of Clinical Nutrition*, 95, pp. 283–9

Makarova et al. (2015): *Journal of the Science of Food and Agriculture*, 95, pp. 560–8

Malik et al. (2016): *American Journal of Epidemiology*, 183, pp. 715–28

Mansoor et al. (2015): *British Journal of Nutrition*, 115, pp. 466–79

Marckmann et al. (2015): *Journal of Renal Nutrition*, 25, pp. 1–5

Maresz (2015): *Integrative Medicine*, 14, pp. 34–8

Markhus et al. (2013): *PLOS One*, 8, p. e7617

Martineau et al. (2017): *British Medical Journal*, 356, p. i6583

Martinez et al. (2014): *Nature Reviews Endocrinology*, 10, pp. 749–60

Martínez-González et al. (2015): *Progress in Cardiovascular Diseases*, 58, pp. 50–60

Martínez Steele et al. (2017): *Public Health Nutrition*, online 16 October

McAfee et al. (2011): *British Journal of Nutrition*, 105, pp. 80–9

McCann & Ames (2009): *American Journal of Clinical Nutrition*, 90, pp. 889–907

McCarty et al. (2009): *Medical Hypothesis*, 72, pp. 125–8

McClain et al. (2013): *Diabetes, Obesity and Metabolism*, 15, pp. 87–90

McDaniel et al. (2011): *Epilepsia*, 52, pp. e7–e11

McGill et al. (2013): *Annals of Medicine*, 45, pp. 467–73

McIsaac et al. (2016): *Annals of the New York Academy of Sciences*, 1363, pp. 155–70

Melkani & Panda (2017): *Journal of Physiology*, 595, pp. 3691–3700

Melnik (2015): *International Journal of Molecular Sciences*, 16, pp. 17048–87

Menendez et al. (2013): *Cell Cycle*, 12, pp. 555–78

Messamore et al. (2017): *Progress in Lipid Research*, 66, pp. 1–13

Metschnikoff (1908): *Beiträge zu einer optimistischen Weltauffassung*. Lehmanns

Michaëlsson et al. (2014): *British Medical Journal*, 349, p. g6015

Michaëlsson et al. (2017): *American Journal of Epidemiology*, 185, pp. 345–61

Michas et al. (2014): *Atherosclerosis*, 234, pp. 320–8

Mollard et al. (2012): *British Journal of Nutrition*, 108, pp. S111–S122

Morris et al. (2016): *JAMA*, 315, pp. 489–97

Mosley & Spencer (2014): *The Fast Diet*. Goldmann

Mozaffarian (2016): *Circulation*, 133, pp. 187–225

Mozaffarian et al. (2011): *NEJM*, 364, pp. 2392–2404

Mozaffarian et al. (2013): *Public Health Nutrition*, 16, pp. 2255–64

Mozaffarian & Rimm (2006): *JAMA*, 296, pp. 1885–99

Müller et al. (2001): *Scandinavian Journal of Rheumatology*, 30, pp. 1–10

Muraki et al. (2013): *British Medical Journal*, 347, p. f5001

Muraki et al. (2016): *Diabetes Care*, 39, pp. 376–84

Nagao & Yanagita (2010): *Pharmacological Research*, 61, pp. 208–12

Nagata et al. (2017): *American Journal of Clinical Nutrition*, 105, pp. 426–31

Niu et al. (2004): *Journal of Biological Chemistry*, 279, pp. 31098–104

O'Donnell et al. (2014): *NEJM*, 371, pp. 612–23

Oh et al. (2010): *Cell*, 142, pp. 687–98

Orlich et al. (2013): *JAMA Internal Medicine*, 173, pp. 1230–8

Orlich & Fraser (2014): *American Journal of Clinical Nutrition*, 100, pp. 353S–358S

Ornish et al. (1990): *Lancet*, 336, pp. 129–33

Ornish et al. (1998): *JAMA*, 280, pp. 2001–7

Osborn & Olefsky (2012): *Nature Medicine*, 18, pp. 363–74

Othmann et al. (2011): *Nutrition Reviews*, 69, pp. 299–309

Parra et al. (2007): *European Journal of Nutrition*, 46, pp. 460–7

Parrella et al. (2013): *Aging Cell*, 12, pp. 257–68

Peou et al. (2016): *Journal of Clinical Lipidology*, 10, pp. 161–71

Perlmutter (2014): *Dumm wie Brot*. Mosaik

Persson et al. (2003): *Food and Chemical Toxicology*, 41, pp. 1587–97

Pietrocola et al. (2014): *Cell Cycle*, 13, pp. 1987–94

Pollak (2012): *Nature Reviews Cancer*, 12, pp. 159–69

Pollan (2011): *64 Grundregeln Essen*. Goldmann

Pottala et al. (2014): *Neurology*, 82, pp. 435–42

Poutahidis et al. (2013): *PLOS One*, 8, p. e68596

Poutahidis et al. (2014): *PLOS One*, 9, p. e84877

Pucciarelli et al. (2012): *Rejuvenation Research*, 15, pp. 590–5

Qi et al. (2013): *Diabetes Care*, 36, pp. 3442–7

Rabenberg et al. (2015): *BMC Public Health*, 15, p. 641

Raji et al. (2014): *American Journal of Preventive Medicine*, 4, pp. 444–51

Ramsden & Domenichiello (2017): *Lancet*, online 29 August

Ranasinghe et al. (2012): *Diabetic Medicine*, 29, pp. 1480–92

Reaven (2005): *Annual Review of Nutrition*, 25, pp. 391–406

Reaven (2012): *Arteriosclerosis, Thrombosis, and Vascular Biology*, 32, pp. 1754–9

Rebello et al. (2014): *Nutrition Journal*, 13, p. 49

van de Rest et al. (2016): *Neurology*, 86, pp. 1–8

Richard et al. (2017): *Journal of Alzheimer's Disease*, 59, pp. 803–14

Richardson et al. (2015): *Experimental Gerontology*, 68, pp. 51–8

Richter et al. (2014): *Journal of Photochemistry and Photobiology B: Biology*, 140, pp. 120–9

Richter et al. (2015): *Advances in Nutrition*, 6, pp. 712–28

Riera & Dillin (2015): *Nature Medicine*, 21, pp. 1400–5

Rigacci et al. (2015): *Oncotarget*, 6, pp. 35344–57

Rizzo et al. (2016): *Nutrients*, 8, p. 767

Roerecke & Rehm (2014): *BMC Medicine*, 12, p. 182

Rosa et al. (2017): *Nutrition Research Reviews*, 30, pp. 82–96

Ross & Bras (1974): *Nature*, 250, pp. 263–5

Ryan & Seeley (2013): *Science*, 339, pp. 918–19

Samuel & Shulman (2016): *Journal of Clinical Investigation*, 126, pp. 12–22

Sanchez et al. (2014): *British Journal of Nutrition*, 111, pp. 1507–19

Santangelo et al. (2016): *Journal of Endocrinological Investigation*, 39, pp. 1295–1301

Santesso et al. (2012): *European Journal of Clinical Nutrition*, 66, pp. 780–8

Santiago et al. (2016): *Nutrition, Metabolism & Cardiovascular Diseases*, 26, pp. 468–75

Saslow et al. (2014): *PLOS One*, 9, p. e91027

Saxton & Sabatini (2017): *Cell*, 168, pp. 960–76

Schenk et al. (2008): *Journal of Clinical Investigation*, 118, pp. 2992–3002

Schmaal et al. (2016): *Molecular Psychiatry*, 21, pp. 806–12

Schröder et al. (2014): *JAMA Internal Medicine*, 174, pp. 1690–2

Schulze et al. (2014): *Molecular Nutrition & Food Research*, 58, pp. 1795–1808

Schwingshackl et al. (2017a): *Advances in Nutrition*, 8, pp. 27–39

Schwingshackl et al. (2017b): *Nutrition & Diabetes*, 7, p. e262

Senftleber et al. (2017): *Nutrients*, 9, p. 42

Sengupta et al. (2006): *Food and Chemical Toxicology*, 44, pp. 1823–9

deShazo et al. (2013): *American Journal of Medicine*, 126, pp. 1018–19

Shen et al. (2015): *Annual Review of Nutrition*, 35, pp. 425–49

Simpson et al. (2003): *Appetite*, 41, pp. 123–40

Simpson et al. (2006): *PNAS*, 103, pp. 4152–6

Simpson & Raubenheimer (2005): *Obesity Reviews*, 6, pp. 133–142

Simpson & Raubenheimer (2012): *The Nature of Nutrition*. Princeton University Press

Simpson & Raubenheimer (2014): *Nature*, 508, p. S66

Siri-Tarino et al. (2015): *Annual Review of Nutrition*, 35, pp. 517–43

Skaldeman (2011): *Lose Weight by Eating*. Little Moon

Skov et al. (1999): *International Journal of Obesity*, 23, pp. 528–36

Sluik et al. (2016): *British Journal of Nutrition*, 115, pp. 1218–25

Smith et al. (2016): *Diabetologia*, online 17 October

Soerensen et al. (2014): *American Journal of Clinical Nutrition*, 99, pp. 984–91

Song et al. (1999): *Mechanisms of Ageing and Development*, 108, pp. 239–51

Song et al. (2016): *JAMA Internal Medicine*, 176, pp. 1453–63

Souza et al. (2008): *American Journal of Clinical Nutrition*, 88, pp. 1–11

Stanford & Goodyear (2014): *Advances in Physiology Education*, 38, pp. 308–14

Stanhope (2015): *Critical Reviews in Clinical Laboratory Sciences*, 53, pp. 52–67

Stanhope et al. (2009): *Journal of Clinical Investigation*, 119, pp. 1322–34

Steven et al. (2013): *Diabetic Medicine*, 30, pp. e135–e138

Steven et al. (2016): *Diabetes Care*, online 21 March

Steven & Taylor (2015): *Diabetic Medicine*, 32, pp. 1149–55

St-Onge et al. (2017): *Circulation*, 135, pp. e96–e121

Strobel et al. (2012): *Lipids in Health and Disease*, 11, p. 144

Stull et al. (2010): *Journal of Nutrition*, 140, pp. 1764–8

Sublette et al. (2006): *American Journal of Psychiatry*, 163, 6, pp. 1100–2

Suez et al. (2014): *Nature*, 514, pp. 181–6

Sultana et al. (2015): *Environmental Monitoring and Assessment*, 187, p. 4101

Sylow et al. (2017): *Nature Reviews Endocrinology*, 13, pp. 133–48

Tang et al. (2014): *Trends in Neurosciences*, 38, pp. 36–44

Tang et al. (2015): *British Journal of Nutrition*, 114, pp. 673–83

Tarasoff-Conway et al. (2015): *Nature Reviews Neurology*, 11, pp. 457–70

Taubes (2008): *Good Calories, Bad Calories*. Anchor Books

Taubes (2011): *Why We Get Fat*. Anchor Books

Taubes (2016): *The Case Against Sugar*. Knopf

Taylor (2013): *Diabetes Care*, 36, pp. 1047–55

Tognon et al. (2017): *American Journal of Clinical Nutrition*, online 10 May

Toledo et al. (2015): *JAMA Internal Medicine*, 175, pp. 1752–60

Toma et al. (2017): *Current Atherosclerosis Reports*, 19, p. 13

Tong et al. (2017): *Nutrients*, 9, p. 63

Törrönen et al. (2012): *American Journal of Clinical Nutrition*, 96, pp. 527–33

Tryon et al. (2015): *Journal of Clinical Endocrinology and Metabolism*, 100, pp. 2239–47

Tuomi et al. (2016): *Cell Metabolism*, 23, pp. 1067–77

Ulven & Holven (2015): *Vascular Health and Risk Management*, 11, pp. 511–24

Van Aller et al. (2011): *Biochemical and Biophysical Research Communications*, 406, pp. 194–99

Verburgh (2015a): *Die Ernährungs-Sanduhr*. Goldmann

Verburgh (2015b): *Veroudering vertragen*. Prometheus/Bert Bakker

Vieira et al. (2016): *PLOS One*, 11, p. e0163044

Vieth (2011): *Best Practice & Research Clinical Endocrinology & Metabolism*, 25, pp. 681–91

Vitaglione et al. (2015): *Critical Reviews in Food Science and Nutrition*, 55, pp. 1808–18

Volek et al. (2008): *Progress in Lipid Research*, 47, pp. 307–18

Wacker & Holick (2013): *Dermato-Endocrinology*, 5, pp. 51–108

Wahrenberg et al. (2005): *British Medical Journal*, 330, pp. 1363–4

Walford (2000): *Beyond the 120-Year Diet*. Four Walls Eight Windows

Wang et al. (2014): *BMC Medicine*, 12, p. 158

Wang et al. (2015): *Journal of the American Heart Association*, 4, p. e001355

Wang et al. (2016a): *Public Health Nutrition*, 19, pp. 893–905

Wang et al. (2016b): *JAMA Internal Medicine*, 176, pp. 1134–45

Weigle et al. (2005): *American Journal of Clinical Nutrition*, 82, pp. 41–8

Wiley (2012): *American Journal of Human Biology*, 24, pp. 130–8

Willcox et al. (2001): *The Okinawa Program*. Three Rivers Press

Willcox et al. (2007): *Annals of the New York Academy of Sciences*, 1114, pp. 434–55

Willcox et al. (2014): *Mechanisms of Aging and Development*, 136–137, pp. 148–62

Willcox & Willcox (2014): *Current Opinion in Clinical Nutrition and Metabolic Care*, 17, pp. 51–8

Willett (2001): *Eat, Drink and Be Healthy*. Free Press

Willett (2006): *Public Health Nutrition*, 9, pp. 105–10

Willett et al. (1995): *American Journal of Clinical Nutrition*, 61, pp. 1402S–1406S

Witte et al. (2014): *Cerebral Cortex*, 24, pp. 3059–68

Wycherley et al. (2012): *American Journal of Clinical Nutrition*, 96, pp. 1281–98

Yang et al. (2014): *JAMA Internal Medicine*, 174, pp. 516–24

Yang & Wang (2016): *Molecules*, 21, p. 1679

Ye et al. (2012): *Journal of Nutrition*, 142, pp. 1304–13

Young & Hopkins (2014): *European Respiratory Review*, 23, pp. 439–49

Zeevi et al. (2015): *Cell*, 163, pp. 1–16

Zhang et al. (2013): *Nature*, 497, pp. 211–16

Zhang et al. (2015): *European Journal of Epidemiology*, 30, pp. 103–13

Zhang et al. (2017): *Journal of Alzheimer's Disease*, 55, pp. 497–507

Zhao et al. (2015): *European Journal of Clinical Nutrition*, pp. 1–7

Zoncu et al. (2011): *Nature Reviews Molecular Cell Biology*, January, pp. 21–35

Zong et al. (2016): *Circulation*, 133, pp. 2370–80

Notes

Introduction

1   Andersen et al. (2012)

2   Described well, for example, in Eenfeldt (2013)

3   For more, see e.g. the analysis published by Dehghan et al. (2017) and the commentary on it by Ramsden & Domenichiello (2017)

4   del Gobbo et al. (2016)

5   Aune et al. (2016)

6   Credit for this analogy must go to the US writer Gary Taubes, with whom I corresponded repeatedly while researching this book. Taubes himself says he borrowed it from a blogger. Taubes has published two books on obesity that are well worth reading: *Why We Get Fat* and his monumental and provocative work *Good Calories, Bad Calories*. Although I remain ultimately unconvinced by the actual dietary advice he provides, his analyses definitely provide food for thought.

7   Assuming that there is a causal relationship — more on this in the rest of the book.

8   Mozaffarian et al. (2011)

9   Esselstyn (2015)

10 Esselstyn (2001), Esselstyn et al. (2014), Esselstyn (2015)

11 Diabetes is a malfunctioning of the body's ability to control blood-sugar levels. Our bodies are constantly trying to keep the level of sugar in our blood stable; not too high and not too low. The ability to regulate that level is lost in patients with diabetes. In that case, there is too much sugar (glucose) circulating in the blood at all times — and not just immediately after a meal, as normal. Whenever I talk about diabetes in this book, I am referring to type-2 diabetes. This is by far the most common form of the disease, which develops gradually, usually later in life, and is heavily dependent on the patient's diet and lifestyle. Being overweight is the main risk factor, and (if necessary, extreme) weight loss and exercise can often not only halt its development, but also even reverse it. Diabetes is defined by a fasting blood-glucose level of at least 126 milligrams per decilitre (i.e. per tenth of a litre, mg/dL) of blood. The main problem with type-2 diabetes is a metabolic disorder known as 'insulin resistance'. When this occurs, the cells of the body, especially those of the muscles and the liver, become increasingly insensitive to the hormone insulin. Insulin is produced by the pancreas and excreted into the blood when we eat something (not only, but especially after a high-carbohydrate meal, which results in high blood-sugar levels). Insulin causes the body's cells to take in the excess sugar from the bloodstream, and there it is used to provide energy, or stored for future use. How can diabetes cause increased blood-sugar levels in patients who have eaten nothing for a number of hours (this is the definition of 'fasting blood-glucose level', which is best measured in the morning, before breakfast)? When we eat nothing for an extended period of time — for example, when we're sleeping at night or fasting — our liver produces glucose,

mainly so that our brain can continue to be supplied with energy at all times. This is why our blood-sugar level remains constant even when we don't eat anything. Insulin inhibits the production of sugar by the liver, since, under normal circumstances, lots of insulin means there is already a sufficient level of glucose in the blood (because, as previously explained, insulin is secreted after a high-carbohydrate meal). However, when the cells of the liver become insulin resistant, they start ignoring that inhibiting effect of insulin. They merrily carry on producing glucose, even when there is already ample sugar in the bloodstream. Thus, 'insulin resistance' leads to an increased fasting blood-glucose level. At the cellular level, one of the causes of insulin resistance is that the cells are malnourished or overnourished and become too fatty. Those fat molecules hinder insulin's signalling pathways inside the cell, causing the cell to take on glucose in healthy patients. The immediate problem with diabetes is that significantly increased blood sugar can cause all sorts of damage due to the tendency of glucose molecules to 'stick' to other molecules — such as protein molecules, for example — which causes, among other things, tissue to become sclerotic (that is, less flexible). This can lead, for example, to the 'calcification' of our veins and arteries, which is also another phenomenon of ageing. The pancreas then attempts to compensate for the cells' resistance to insulin by increasing insulin production and excretion, which is also damaging. For these reasons, diabetes is not just any old illness, but could be described as the ultimate *disease-causing* disease. In this way, diabetes is similar to the ageing process, as ageing also increases the risk of developing practically all chronic conditions (diabetes appears to accelerate at least some ageing processes). It is the leading cause of blindness in adulthood, as well as causing kidney failure and foot and leg amputations. It also drastically increases patients' risk of developing cardiovascular disease and cancer. Long-term overproduction of insulin also puts a strain on the pancreas, which is then in danger of giving up out of 'exhaustion' — at this stage, the body produces *too little* insulin. The symptoms associated with diabetes include excessive thirst and the associated frequent urination as the body tries to rid itself of the extra sugar by flushing it out. As a result, diabetic patients' urine smells and tastes sweet — hence the official name of the disease, 'diabetes mellitus', which comes from the Greek for 'discharge' or 'passer-through' and the Latin for 'sweet like honey'.

12   Lim et al. (2011), Steven et al. (2013), Steven & Taylor (2015), Steven et al. (2016)

13   Bredesen (2014), Bredesen et al. (2016), see also Bredesen (2017)

14   Bredesen et al. (2016)

15   This is cautiously worded, as the results are still new (and, many will say, too good to be true). So a healthy dose of scepticism is appropriate here, until other research groups are (hopefully) able to replicate these results.

16   German Federal Statistical Office, figures for 2014

17   Fries (1980), Fries et al. (2011)

18   Couzin-Frankel (2014), Solon-Biet et al. (2014)

19   Levine et al. (2014)

Chapter 1. Proteins I

1   Simpson et al. (2006), personal communication

2   Simpson & Raubenheimer (2012)

3   With thanks to Stephen Simpson for the video material

4   This should really read 'kilocalories' (= 1,000 calories), but the simpler word calorie has become so established in normal parlance that I decided to use it in this book for ease of reading. But please bear in mind, that when I speak of a 'calorie', I actually

mean a 'kilocalorie', also called a 'large calorie', 'food calorie', or 'kilogram calorie'.

5   Simpson et al. (2003), Simpson et al. (2006), Simpson & Raubenheimer (2012)

6   Nationale Verzehrsstudie II (2008), to be found at: https://www.bmel.de/
    DE/Ernaehrung/GesundeErnaehrung/_Texte/NationaleVerzehrsstudie_
    Zusammenfassung.html

7   Simpson & Raubenheimer (2012)

8   Simpson & Raubenheimer (2014)

9   Simpson & Raubenheimer (2005), Gosby et al. (2014)

10 Gosby et al. (2011)

11 deShazo et al. (2013)

12 Data from Martínez Steele et al. (2017)

13 Pollan (2011)

14 Daley et al. (2010), McAfee et al. (2011)

15 Due et al. (2004), Skov et al. (1999)

16 Weigle et al. (2005)

17 Santesso et al. (2012). For more reviews and meta-analyses that reach the same
    conclusion, see Leidy et al. (2015), Mansoor et al. (2015), Martinez et al. (2014),
    Clifton et al. (2014), Wycherley et al. (2012)

Chapter 2. Proteins II

1   https://www.wsj.com/articles/SB107637899384525268

2   Heart failure is a serious condition, but it does not actually mean that a patient's heart
    has failed. It means that the heart is not able to pump blood around the body very
    efficiently.

3   http://edition.cnn.com/2002/HEALTH/diet.fitness/04/25/atkins.diet/

4   Souza et al. (2008)

5   Gardner (2012)

6   Johnston et al. (2014)

7   Alhassan et al. (2008), Dansinger et al. (2005)

8   Ross & Bras (1974)

9   Lagiou et al. (2012), Marckmann et al. (2015)

10 Ross & Bras (1974)

11 https://www.elsevier.com/connect/controlling-protein-intake-may-be-key-to-longevity

12 Levine et al. (2014)

13 Freedman et al. (2012)

14 Folkman & Kalluri (2004)

15 Levine et al. (2014)

16 Zoncu et al. (2011), Laplante & Sabatini (2012), Johnson et al. (2013), Saxton &
    Sabatini (2017)

17 Hosios et al. (2016)

18 Parrella et al. (2013)

19 Hosios et al. (2016)

20 Pietrocola et al. (2014)

21 Van Aller et al. (2011)

22 Levine et al. (2014)

23 Fry et al. (2011)

24 Dennison et al. (2017)

25 Levine et al. (2014)

26 Song et al. (2016)

27 Ross & Bras (1974)

28 Lee et al. (2008)

29 Konner & Eaton (2010)

30 This argument is taken from the work of the eccentric gerontologist Roy Walford (2000)

31 Wang et al. (2016a)

32 Larsson & Orsini (2014)

33 Since I have now broached the subject, here's a note to those who expressly advocate eating generous amounts of meat (typically champions of the Paleo and low-carb diets): in general, I welcome maverick opinions and attempts to question the mainstream. It's certainly stimulating and beneficial from a purely academic and intellectual point of view. In this case, however, there is another dimension at play. Such 'polemical' arguments contradict the findings of most empirical studies, and are therefore probably factually incorrect. It's *one* thing to accept the suffering of animals, because — unfortunately — almost all research results show that the welfare of the human race depends on it ('According to the latest science, meat is so extremely good for our health that I advocate extremely high meat consumption despite the high moral cost we have to pay for it'). It's another thing to take this morally ambivalent stance *despite* the fact that almost all the scientific research indicates that eating a lot of meat is *not* good for our health. What you argue for, on very shaky ground, could cost many people their health and cause the suffering of countless animals. In other words, although there is overwhelming scientific evidence in favour of the far more ethically sound position (less meat means less disease, see figure 6.2 in chapter 6), that position is rejected in favour of one that results in more suffering for humans and animals.

34 The farm is near Ochsenfurt: https://www.bayrischer-feenhof.de/

35 Be sure to check the sugar content of peanut butter, which can vary greatly between brands. Peanut butter often also contains palm oil, the effects of which have not been thoroughly investigated. Some kinds of peanut butter may contain trans fats, which are unequivocally harmful. For more on the subject of sugar, see chapter 4; for more on trans fats, turn to chapter 7; and information about palm oil is found in chapter 8. It is not too difficult to find peanut butter that does not contain added sugar, palm oil, or trans fats.

36 Richter et al. (2015), Malik et al. (2016)

37 Song et al. (2016)

38 Richter et al. (2015)

39 Parrella et al. (2013)

40 Fontana & Partridge (2015), Ables et al. (2016), McIsaac et al. (2016)

41 Cavuoto & Fenech (2012) give a summary of the methionine content of various foodstuffs

42 McCarty et al. (2009)

43 Cavuoto & Fenech (2012), and there are similar debates about the effects of other amino acids, see e.g. Maddocks et al. (2017)

44 For a specific review of kefir, see Rosa et al. (2017). As early as 1907, the Russian immunologist and Nobel Prize laureate Ilya Ilyich Mechnikov speculated in his work *The Prolongation of Life: Optimistic Studies* about the healing powers of 'soured milk'. One of the book's central topics is healthy ageing. Mechnikov recognised the beneficial effect of lactic-acid bacteria on our gut and microbiome (the billions of bacteria that populate the gut). He wrote, 'For more than eight years I took, as a regular part of my diet, soured milk at first prepared from boiled milk, inoculated with a lactic leaven ... I am very well pleased with the result ... If it be true that our precocious and unhappy old age is due to poisoning of the tissues (the greater part of the poison coming from the large intestine inhabited by numberless microbes), it is clear that agents which arrest intestinal putrefaction must at the same time postpone and ameliorate old age. This theoretical view is confirmed by the facts regarding races which live chiefly on soured milk, and amongst which great ages are common.'

45 Poutahidis et al. (2013)

46 ibid.

47 ibid.

48 Schenk et al. (2008)

49 Poutahidis et al. (2013), Levkovich et al. (2013), Poutahidis et al. (2014)

50 Santiago et al. (2016)

51 Sanchez et al. (2014)

52 ibid.

53 Lee et al. (2015a)

54 ibid.

55 Willcox et al. (2014), Willcox et al. (2001)

56 Fraser and Shavlik (2001)

57 Orlich et al. (2013), Orlich & Fraser (2014)

58 Buettner (2015)

59 Burr et al. (1989), Parra et al. (2007)

60 Zhao et al. (2015)

61 ibid.

62 Belin et al. (2011)

63 Bellavia et al. (2017)

64 Gil & Gil (2015), Mozaffarian & Rimm (2006)

65 Morris et al. (2016)

66 Gu et al. (2015), Raji et al. (2014)

67 van de Rest et al. (2016)

68 Gil & Gil (2015)

69 This is a reference in particular to pulses, which provide not only a large amount of

protein, but also dietary fibre, which also makes us feel full faster. Experiments have shown that pulses such as beans and peas are better at making us feel full than meat; see also the study by Kristensen et al. (2016). More on this in chapter 6.

70 cf. the excellent review by Mozaffarian (2016)

Chapter 3. Intermezzo

1 Atherosclerosis is a hardening of an artery specifically due to a build-up of plaque (from Ancient Greek *athera*, meaning 'gruel', and *sklerosis*, meaning 'hardening'), and can be easily confused with arteriosclerosis and arteriolosclerosis. 'Arteriosclerosis' is a general term describing any hardening of medium or large arteries (from Greek *arteria*, meaning 'artery'); arteriolosclerosis is any hardening of arterioles (small arteries).

2 Okinawa: Willcox et al. (2007), Willcox & Willcox (2014), Willcox et al. (2014); figures for Germany: personal communication with the Federal Statistical Office. At the end of 2014, there were 17,474 centenarians in Germany, among a population of 81,197,537.

3 Fontana et al. (2010)

4 Kaplan et al. (2017)

5 Nationale Verzehrsstudie II (2008), to be found at: https://www.bmel.de/ DE/Ernaehrung/GesundeErnaehrung/_Texte/NationaleVerzehrsstudie_ Zusammenfassung.html

6 https://www.dge.de/presse/pm/kohlenhydrate-in-der-ernaehrung/

7 See, for example the perhaps well-meant but naive and tendentious documentary film *What the Health* (2017): https://www.whatthehealthfilm.com/

8 Willett et al. (1995), Willett (2006)

9 Okinawa: Willcox et al (2014); Tsimané: Kaplan et al. (2017); Adventists: Orlich & Fraser (2014), Mediterranean diet: Souza et al. (2008)

10 Shen et al. (2015)

11 Martínez-González et al. (2012)

12 Martínez-González et al. (2012), Martínez-González et al. (2015), Schröder et al. (2014)

13 Martínez-González et al. (2015)

14 Estruch et al. (2013)

15 de Lorgeril et al. (1994)

16 Luciano et al. (2017)

17 Jacka et al. (2017)

18 Appel & Van Horn (2013)

Chapter 4. Carbohydrates I

1 Gary Taubes cites this quote from Lewis Cantley in his highly recommendable article 'Is Sugar Toxic?' in *The New York Times Magazine*, to be found online at: https:// www.nytimes.com/2011/04/17/magazine/mag-17Sugar-t.html. I contacted Cantley personally, and he not only affirmed this statement, but also provided me with a detailed explanation of why he believes it's justified (email dated 14 July 2016). This chapter takes into account parts of that justification.

2   Tryon et al. (2015)

3   Cantley (2014)

4   Johnson (2015)

5   Email dated 14 July 2016; on the addictive character of sugar, see also Ahmed et al. (2013)

6   Cantley (2014)

7   https://www.youtube.com/watch?v=CUOu3ELNVxc

8   Willcox et al. (2001)

9   Fetissov (2017)

10 Suez et al. (2014)

11 Grassi et al. (2008)

12 A truly wonderful green tea from Japan containing large amounts of the beneficial plant substance epigallocatechin gallate (EGCG). I owe this tip to the book *Foods That Fight Cancer* by the two Quebecois research scientists Richard Béliveau and Denis Gingras.

13 Törrönen et al. (2012)

14 Fardet (2015)

15 Muraki et al. (2013)

16 Taubes (2008)

17 Email dated 14 July 2016. The protein Cantley discovered is called PI3K (phosphoinositide 3-kinase). A good introduction for those interested in learning more about the link between insulin, PI3K, and cancer can be found in the summary by Pollak (2012).

18 For more on insulin resistance and its pathological consequences, see also Reaven (2012)

19 Yang et al. (2014)

20 Leung et al. (2014), see also Lee et al. (2015b)

21 My accounts are largely based on Taubes (2008), Siri-Tarino et al. (2015), Stanhope (2015), Cantley (2014), Lyssiotis & Cantley (2013), Herman & Samuel (2016), Lustig (2016)

22 Maersk et al. (2012)

23 Stanhope et al. (2009); Aeberli et al. (2011, 2013) reach similar conclusions.

Chapter 5. Carbohydrates II

1   His upper, 'systolic' value was more than 200 mmHg

2   Email dated 16 September 2016

3   Personal communication with Sten Sture Skaldeman. See also Skaldeman (2011) and Eenfeldt (2013), as well as Skaldeman's website, http://www.skaldeman.se/

4   Gardner et al. (2007), Gardner (2012)

5   Gardner (2012), see also Hjorth et al. (2017)

6   Gardner (2012)

7   McClain et al. (2013)

8   Samuel & Shulman (2016)

9   In this case 'I' really means my wife, Sina Bartfeld

10   Farin et al. (2006)

11   Wahrenberg et al. (2005)

12   People who carry certain genes that increase the risk of type-2 diabetes have been shown to do better on a fat-heavy diet — see, for example, Qi et al. (2013). This is hardly surprising, as one of the central problems in type-2 diabetes is insulin resistance.

13   Many people might see a connection between these indicators and 'metabolic syndrome', and they are, of course, correct. Since it is now known that insulin resistance lies at the core of that condition, it is now sometimes also known as insulin-resistance syndrome. For more, see the brief review by Gerald Reaven (2005) of Stanford University, who contributed significantly to the discovery of this syndrome.

14   Costello et al. (2016), Ranasinghe et al. (2012)

15   Bao et al. (2009)

16   An experiment by Kanarek & Ho (1984) firstly involved making rats diabetic and then observing what happened when they fed the diabetic animals more carbohydrates. As their bodies were no longer able to process this fuel properly, they initially attempted to compensate for the energy deficit by eating more. When that didn't help, the rats changed their strategy and after three weeks they switched to eating more fat. It was as if the rats had gradually found out by trial and error that their bodies could cope better with fat!

17   Volek et al. (2008), see also Esposito et al. (2009) and Saslow et al. (2014)

18   Sylow et al. (2017)

19   Catenacci & Wyatt (2007), Chin et al. (2016)

20   Blundell et al. (2009)

21   Caudwell et al. (2009)

22   Knowler et al. (2002), Stanford & Goodyear (2014), Smith et al. (2016)

23   Santangelo et al. (2016), see also Schwingshackl et al. (2017)

24   Han et al. (2007), Nagao & Yanagita (2010)

25   Ghorbani et al. (2014), see also the spectacular findings of Chuengsamarn et al. (2012)

26   Stull et al. (2010)

27   Escarpa & González (1998)

28   Schulze et al. (2014), Makarova et al. (2015)

29   Liu et al. (2013)

30   Grassi et al. (2005)

31   Grepner et al. (2015)

32   Lim et al. (2011), Steven et al. (2013), Steven & Taylor (2015), Steven et al. (2016)

Chapter 6. Carbohyrdates III

1 Kamiloglu et al. (2014)

2 Mozaffarian et al. (2013)

3 Linseed is unbeatable, from a purely fibre point of view: 100 grams of linseeds contain no less than 39 grams of fibre, with no other carbohydrates (the rest is made up mainly of beneficial fats)!

4 Verburgh (2015)

5 Fardet (2010)

6 This is how I bake my favourite sourdough bread: the basic ingredients you will need are — two 75-gram packs of liquid sourdough starter (available in supermarkets, organic food stores, and online shops). Some people recommend using 75 grams of liquid sourdough starter per 500 grams of flour, but I like my sourdough bread super-sour and use twice that amount. For the flour, I recommend 300 grams of wholegrain rye flour, plus 200 grams of wholegrain wheat flour. A packet of dry yeast (10 grams) or, even better, fresh yeast (approx. 20 grams). Approx. 2 teaspoons of salt. A good 400 mL of lukewarm water. Method: pour the lukewarm water into a plastic mixing bowl, add the yeast (you can also add a small spoonful of honey and a pinch of salt to enrich it). Mix together and briefly leave to soak. Add the sourdough and mix again. Now add the flour. I also add some linseeds and/or chia seeds; some wheatgerm, roughly chopped nuts, or whole rye grains are also delicious. Don't forget to add salt, and, finally, add some rapeseed or olive oil. Mix well (with the dough hook on your mixer, as it gets very sticky). Cover the bowl with a cloth and leave to rise in a warm place (for example, in the oven at approx. 50 degrees) for at least 30 minutes. The dough should rise in this time. Knead well again (the dough will lose its volume again). Place the dough in a silicone baking mould — not too full as the dough will expand a lot again. Sieve a little flour over the top of the loaf, just to make it look even more appetising. Again, let it rise in a warm place for a further hour. Preheat the oven, ideally one with a grill function, to 275 degrees. Then give the loaf a blast of heat for as long as possible so that it forms a crunchy crust, but without letting it burn, of course. In my experience, that should not be longer than about 30 minutes. Then reduce the heat to 200 degrees and finish baking for about ten minutes (so that the total baking time is around 40 minutes). When it's baked, remove the loaf and leave it to cool on a wire rack — this is important as the bread will 'sweat' after it leaves the oven. Best enjoyed fresh from the oven! I'm always experimenting with this basic recipe. For example, I recently added around 80 grams of linseed flour (and a corresponding amount of extra water) to make the bread even richer in proteins and fibre. It turned out to be tasty, too …

7 This theory was first suggested by the US biochemist Bruce Ames. See, for example, Ames (2005)

8 Othman et al. (2011), Rebello et al. (2014), Hollænder et al. (2015)

9 Desai et al. (2016)

10 Koh et al. (2016)

11 Review by Koh et al. (2016), Fetissov (2017)

12 Aune et al. (2016) and Zong et al. (2016) came to a similar conclusion; see also Ye et al. (2012), which includes a meta-analysis of the results of 21 studies

13 Johnson (2015)

14 Davis (2003). His colleague David Perlmutter takes a similar view, writing in his book *Grain Brain*, 'Modern grains are silently destroying your brain. By "modern", I'm not just referring to the refined white flours, pastas, and rice that have already been

demonised by the anti-obesity folks; I'm referring to all the grains that so many of us have embraced as being healthful — whole wheat, whole grain, multigrain, seven-grain, live grain, stone-ground, and so on. Basically, I am calling what is arguably our most beloved dietary staple a terrorist group that bullies our most precious organ, the brain.'

15   Lebwohl et al. (2015), Hadjivassiliou et al. (2014)

16   Reviews are qualitative summaries (critically examining the content of various studies and summarising them to form an overall picture). Meta-analyses are quantitative summaries (gathering the data from several studies together and subjecting them to a fresh statistical evaluation).

17   Fardet & Boirie (2014)

18   ibid.

19   An example of someone who does argue this is the friendly vegan Michael Greger, in his very worthwhile book *How Not to Die* (2015) and elsewhere. His website, nutritionfacts.org, is also worth a visit!

20   As mentioned previously, if there really is a causal effect here — it is not, or not yet, proven by these studies.

21   Email dated 13 August 2015

22   Zeevi et al. (2015)

23   My thanks to Jennie Brand-Miller (also known as 'GI Jennie') of the University of Sydney for providing this data

24   Zeevi et al. (2015), see also this interesting talk by the Israeli biologist, Eran Segal: https://www.youtube.com/watch?v=0z03xkwFbw4&feature=youtu.be

25   Atkinson et al. (2008), Brand-Miller et al. (2010), Goletzke et al. (2016), Sluik et al. (2016)

26   Brand-Miller et al. (2009)

27   Muraki et al. (2016)

28   Brand-Miller et al. (2010)

29   McGill et al. (2013)

30   Fardet & Boirie (2013)

31   Brand-Miller et al. (2010) give it a GI of 48

32   Sultana et al. (2015)

33   Hojsak et al. (2015), Karagas et al. (2016), see also https://www.bfr.bund.de/en/frequently_asked_questions_on_arsenic_levels_in_rice_and_rice_products-194425.html

34   Sengupta et al. (2006), see also Carey et al. (2015)

35   The cookery book *Vegan for Fit* by Attila Hildmann contains a recipe for lentils with vegetables and lime-and-sunflower-seed pesto that I can't recommend highly enough (it's amazing)

36   Kim et al. (2016)

37   Jenkins et al. (2012)

38   Young & Hopkins (2014)

39   Hermsdorff et al. (2011), see also Mollard et al. (2012)

40   Buettner (2015)

41   This is the recipe for hummus I got from my university friend Christian Keyser: the secret to its success is good sesame paste (tahini). This is how to make it: take six tablespoons of tahini, 350 grams of soaked chickpeas, the juice of one lemon, one to two cloves of garlic, a pinch of salt, caraway, and maybe some ras el hanout (an exotic spice mix) if you have it. Blitz all the ingredients thoroughly in a mixer and finish off with a glug of olive oil.

42   Which is why the title they gave their paper was 'Legumes: the most important dietary predictor of survival in older people of different ethnicities'; see Darmadi-Blackberry et al. (2004)

Chapter 7. Intermezzo

1   Here is a recent example, which was circulated with little criticism by the media: Guo et al. (2017)

2   https://www.foodpolitics.com/2016/03/six-industry-funded-studies-the-score-for-the-year-15612/

3   Lesser et al. (2007)

4   Michaëlsson et al. (2014), Tognon et al. (2017)

5   Melnik (2015)

6   Wiley (2012), Melnik (2015), Harrison et al. (2017)

7   Bayless et al. (2017)

8   Ji et al. (2014)

9   Tognon et al. (2017)

10   Michaëlsson et al. (2014)

11   Song et al. (1999)

12   Song et al. (1999), Michaëlsson et al. (2014)

13   Michaëlsson et al. (2017)

14   Wang et al. (2014)

15   Michaëlsson et al. (2014)

16   Willett (2001)

17   Crippa et al. (2014), see also Je & Giovannucci (2014) and Gunter et al. (2017)

18   Grosso et al. (2017)

19   Crippa et al. (2014), see also Je & Giovannucci (2014) and Gunter et al. (2017)

20   Takahashi et al. (2017), Pietrocola et al. (2014)

21   Furman et al. (2017)

22   Cai et al. (2012), Rebello & Van Dam (2013)

23   Crioni et al. (2015)

24   Grosso et al. (2017)

25   Rhee et al. (2015)

26   Tang et al. (2015), see also Zhang et al. (2015)

27   Yang & Wang (2016)

28   Van Aller et al. (2011)

29   Bettuzzi et al. (2006)

30 Guo et al. (2017)

31 Costanzo et al. (2011), Roerecke & Rehm (2014), Toma et al. (2017)

32 Bell et al. (2017)

33 Brien et al. (2011), Gepner et al. (2015), Holst et al. (2017)

34 Richard et al. (2017)

35 Di Castelnuovo et al. (2006)

36 Bellavia et al. (2014)

37 Bellavia et al. (2014), Knott et al. (2015)

38 Bagnardi et al. (2014)

39 Cao et al. (2015)

40 Halsted et al. (2002), Chen et al. (2014), de Batlle et al. (2015)

41 All this points towards a drinking culture typical of many parts of the Mediterranean (Greece, Italy, Spain), which has been found to be very healthy and which is associated with a reduced risk of mortality; see Gea et al. (2014), see also Bagnardi et al. (2008) and Vieira et al. (2016)

Chapter 8. Fats I

1   https://en.wikipedia.org/wiki/Moai

2   Harrison et al. (2009)

3   *Science*, 362, pp. 1602–3, 2009

4   de Cabo et al. (2014), Richardson et al. (2015)

5   Finkel (2015), Madeo et al. (2015)

6   Kennedy & Lamming (2016)

7   Riera & Dillin (2015)

8   McDaniel et al. (2011)

9   Toledo et al. (2015)

10 ibid. My thanks go to Estefanía Toledo and Miguel Martínez-González of the University of Navarra for the raw data.

11 Esselstyn (2015)

12 Ornish et al. (1990), Ornish et al. (1998)

13 Esselstyn (2015)

14 Without trans fats, palm oil, added sugar, or salt

15 Guasch-Ferré et al. (2017)

16 Wang et al. (2015), see also the meta-analysis by Peou et al. (2016)

17 Grosso et al. (2015)

18 Grosso et al. (2015), assuming there is a causal relationship here

19 Nagao & Yanagita (2010)

20 For reviews see e.g. Michas et al. (2014) and Calder (2015)

21 Willett (2001)

22 Kavanagh et al. (2007)

NOTES

23 Wang et al. (2016b)

24 Dehghan et al. (2017)

25 Wang et al. (2016b)

26 Vitaglione et al. (2015)

27 Beauchamp et al. (2005)

28 Khanfar et al. (2015)

29 Rigacci et al. (2015)

30 This theory is known as 'xenohormesis'; for more in general, see Howitz & Sinclair (2008); for more specifically on olive oil, see Menendez et al. (2013)

31 Latreille et al. (2012)

32 Joven et al. (2014)

33 Casal et al. (2010)

34 Persson et al. (2003)

35 For a critical comparison with olive oil, see Hoffman & Gerber (2014)

Chapter 9. Fats II

1 Rosqvist et al. (2014)

2 Mancini et al. (2015)

3 For more information see, among others, the study carried out by the WWF called *Palm Oil Report Germany: Searching for Alternatives*: https://mobil.wwf.de/fileadmin/fm-wwf/Publikationen-PDF/WWF_Report_Palm_Oil_-_Searching_for_Alternatives.pdf

4 https://www.dge.de/wissenschaft/weitere-publikationen/fachinformationen/trans-fettsaeuren/

5 Bjermo et al. (2012)

6 Pimpin et al. (2016)

7 Siri-Tarino et al. (2015), de Goede et al. (2015)

8 Soerensen et al. (2014)

9 More precisely: vitamin K2, also known as 'menaquinone'

10 Maresz (2015)

11 Li et al. (2012), Anderson et al. (2016)

12 Juanola-Falgarona et al. (2014)

13 McCann & Ames (2009), see also this talk by the US biochemist Bruce Ames (who has come up with an interesting theory that's supported by a series of scientific findings): https://www.youtube.com/watch?v=ZVQmPVBjubw

14 Nagata et al. (2017)

15 Pucciarelli et al. (2012), see also de Cabo et al. (2014)

16 Esatbeyoglu et al. (2016)

17 Eisenberg et al. (2009)

18 Email from Frank Madeo of the University of Graz, dated 8 February 2017

19 Eisenberg et al. (2016)

267

20   Ali et al. (2011)

21   Esatbeyoglu et al. (2016)

22   A meta-analysis by Tong et al. (2017) yielded a neutral result concerning the link to overall risk of mortality; other large-scale reviews like that of Siri-Tarino et al. (2015) give a more positive picture.

23   For a review which looks in detail at the issue of replacing foods in our diet, see Siri-Tarino et al. (2015)

Chapter 10. Fats III

1   https://www.fischinfo.de/index.php/markt/datenfakten/4856-marktanteile-2016

2   Strobel et al. (2012)

3   https://www.fischinfo.de/index.php/markt/datenfakten/4856-marktanteile-2016

4   Strobel et al. (2012), Cladis et al. (2014), Henriques et al. (2014)

5   Guimarães et al. (2015)

6   https://www.daserste.de/information/wissen-kultur/w-wie-wissen/sendung/2011/die-pangasius-luege-100.html, or see here to gain an impression https://www.youtube.com/watch?v=Px9Enx74kjA

7   Cladis et al. (2014)

8   Niu et al. (2004), see also Calder (2016)

9   Calder (2016)

10   Witte et al. (2014)

11   For a review, see Messamore et al. (2017)

12   Sublette et al. (2006)

13   Chhetry et al. (2016)

14   Schmaal et al. (2016)

15   For a recent study on this, see Zhang et al. (2017), see also Witte et al. (2104), Pottala et al. (2014), and the review by Messamore et al. (2017)

16   Hibbeln (2002), Markhus et al. (2013)

17   Hibbeln (2002)

18   Oh et al. (2010)

19   Zhang et al. (2013)

20   In this case, using linseed oil, which is made up mainly of alpha-linolenic acid

21   Cintra et al. (2012)

22   Bender et al. (2014)

23   Bell et al. (2014), Cardoso et al. (2016)

24   https://www.vitalstudy.org/

25   Bell et al. (2014), Chen et al. (2016)

26   Giuseppe et al. (2014a), Giuseppe et al. (2014b), Senftleber et al. (2017)

27   Bell et al. (2014)

28   Ulven & Holven (2015)

29   Eyres et al. (2016)

Chapter 11. No vitamin pills

1 Schwingshackl et al. (2017)

2 For reviews see, for example, Wacker & Holick (2013) and Holick (2017)

3 Holick (2017)

4 Holick et al. (2011), Vieth (2011)

5 Richter et al. (2014), Rabenberg et al. (2015)

6 Richter et al. (2014). Note: annoyingly, two different units of measurement are customarily used for this metric: nanomoles per litre (nmol/L) and nanograms per millilitre (ng/mL). 1 ng/mL is equivalent to 2.5 nmol/L. That means 50 nmol/L is the same as 20 ng/mL, 75 nmol/L equals 30 ng/mL. The presumed ideal concentration is 75 nmol/L, or 30 ng/mL, or above. Weight and 'mole' are two different metrics. While weight is a measure of how heavy something is, a mole is a unit showing how many particles of a substance are present.

7 Martineau et al. (2017)

8 Bjelakovic et al. (2014)

9 Chowdhury et al. (2014)

10 For newborn babies: 400 units (10 micrograms) per day, ideally immediately from birth (especially if the baby is breastfed, as mothers often don't have enough vitamin D themselves and their milk is therefore deficient in vitamin D). A daily dose of 600 units (15 micrograms) is generally recommended for children older than one year.

11 Rabenberg et al. (2015)

12 Rizzo et al. (2016)

13 Greger (2015)

14 Green et al. (2017)

15 Kennedy (2016), Green et al. (2017)

16 Huang et al. (2012)

17 Douaud et al. (2013)

18 Jernerén et al. (2015)

19 https://www.vitalstudy.org/

Chapter 12. Timing your eating

1 Hatori et al. (2012), Chaix et al. (2014)

2 Which is indeed what happens to mice who feed only within a limited time window, as explained to me by the leader of the research team at the Salk Institute, Satchidananda Panda (email dated 26 April 2017). However, time-restricted feeding also leads to changes in the gut microbiome in such a way as to make the body unable to absorb certain carbohydrates. By the way, Panda personally tries to eat within a time window of ten to 11 hours. If he is trying to lose a little weight, he restricts the time window further, to six to eight hours.

3 Melkani & Panda (2017)

4 Hutchison et al. (2017)

5 Jakubowicz et al. (2013)

6 St-Onge et al. (2017), see also Kahleova et al. (2017)

7   Tuomi et al. (2016)

8   cf. Kessler et al. (2007)

9   Gill & Panda (2015)

10   Longo & Panda (2016)

11   Tarasoff-Conway et al. (2015)

12   Taubes (2008)

13   Lim et al. (2011), Taylor (2013)

14   Goldhamer et al. (2002)

15   Müller et al. (2001)

16   For more on this topic, see the popular book by Mosley & Spencer (2014)

17   St-Onge et al. (2017)

Epilogue

1   For more details, see Nationale Verzehrsstudie II (2008), to be found at: https://www.bmel.de/DE/Ernaehrung/GesundeErnaehrung/_Texte/NationaleVerzehrsstudie_Zusammenfassung.html

2   Here once again, opinions differ. A moderate consumption of approx. one to a maximum of two teaspoons a day appears to be the optimum. Remember that many types of food — from bread and sausages to preserved olives, etc. — often contain a lot of salt, making it easy to eat too much. I think it makes sense to use it cautiously. See Graudal et al. (2014), O'Donnell et al. (2014)

Figure credits

Sina Bartfeld: figs. 6.1, 8.2, molecule diagrams

Sina Bartfeld/Bas Kast: compass needles

Bas Kast/Inka Hagen: fig. 5.1

Bas Kast: figs. 0.1, 0.3, 0.4, 2.1, 2.3, 2.5, 2.6, diet pie charts, 4.2, 6.2, 6.3, 7.1, 8.1, 8.2, 8.3, 9.1, 10.1, 10.2, 10.3, 11.1

Stephen Simpson/University of Sydney: fig. 1.1

Wikimedia Commons/Public Domain: fig. 4.1

Figures were taken from the following publications:

C. B. Esselstyn: 'Resolving the Coronary Artery Disease Epidemic Through Plant-Based Nutrition', in: *Preventive Cardiology* 4, No. 4 (2001): pp. 171–7, Fig. 1 © 2001, Preventive Cardiology: fig 0.2

T. Poutahidis et. al.: 'Microbial reprogramming inhibits Western diet-associated obesity', in: *PLOS One* 8, No. 7 (2013): e68596. doi: 10.1371/journal.pone.0068596. Print 2013, Fig. 2a: fig. 2.2

D. E. Lee, C. S. Huh, J. Ra, I. D. Choi, J. W. Jeong, S. H. Kim, J. H. Ryu et al.: 'Clinical Evidence of Effects of Lactobacillus plantarum HY7714 on Skin Aging: A Randomized, Double Blind, Placebo-Controlled Study', in: *Journal of Microbiology and Biotechnology* 25, No. 12 (2015), pp. 2160–8. doi: 10.4014/jmb.1509.09021. Print 2015, Fig. 4 © 2015, The Korean Society For Microbiology And Biotechnology: fig. 2.4

M. Hatori, C. Vollmers, A. Zarrinpar, L. DiTacchio et al.: 'Time-restricted feeding without reducing caloric intake prevents metabolic diseases in mice fed a high-fat diet', in: *Cell Metabolism* 15, No. 6 (2012), pp. 848–60, Fig. 1j © 2012, Rights Managed by Elsevier/Copyright Clearance Center: ch. 12 mice

Index